CLAUDE CODE FOR NON-DEVELOPERS

The Non-Technical Field Guide to Building Real Apps With AI

MODERNMIND PUBLICATIONS

Copyright

Contents

1
What Claude Code Is

In February 2026, Anthropic ran a hackathon, which is a timed competition where builders create working software from scratch in a set number of days. They called it "Built with Opus 4.6" and opened it to anyone. More than 13,000 people applied. Anthropic selected 500 builders, gave them $100,000 in Claude API credits to split, and set them loose for one week. Six judges from Anthropic assessed the results, scoring each project on how creative the technical approach was, the quality of the actual implementation, and what kind of impact the tool might have if it shipped.

Three of the five winners weren't software developers.

There wasn't a separate category for non-technical builders. They competed on the same playing field as everyone else and beat hundreds of experienced developers to win.

One of the winners is Dr. Michal Nedoszytko, an interventional cardiologist in Brussels. He treats heart patients for a living, which means software development is something he does on the side. He'd noticed a gap between what happens during a cardiology visit and what patients actually remember after they leave. Visit notes get filed away while the discharge instructions and follow-up recommendations get forgotten before the patient even gets home. So Dr. Nedoszytko built Postvisit, a tool that takes visit transcripts and medical records and turns them into personalized, ongoing health guidance for each patient. He took third place. Postvisit was real software that addressed a problem he understood better than any software developer could, because he'd watched it happen thousands of times in his own practice.

Another participant, a California attorney, built CrossBeam, a tool that speeds up the building permit process by automating code compliance checks and plan reviews. Anyone who's worked with California's permit system can guess why someone would want to fix it. He'd been working through that system for clients every day, and he knew exactly what a working tool should do. He didn't need to know any JavaScript. Claude Code handled all of that.

A musician built a generative music system. The interface is a MIDI controller, the kind keyboard players and producers use. You play chords on the controller and Claude directs a four-track generative band fast enough to keep up with you live.

An infrastructure worker from Uganda built a tool that turns road footage into investment recommendations for transportation projects. The system he built analyzes that footage and turns it into actionable data for the people making transportation decisions.

What all of them had in common was that they witnessed the problem every day. Knowing what needed to exist was the hard part, and they had that. Claude Code took care of the technical execution. Domain expertise turned out to matter more than programming knowledge, at least in this competition.

People who build software without traditional programming backgrounds make up the emerging majority of Claude Code users now. The hackathon results are a preview of what the user base actually looks like in 2026.

WHAT CODING WITH AI USED TO LOOK LIKE

None of those hackathon results would have been possible even a year earlier. Anyone who tried building something with AI before tools like Claude Code existed already knows why. The setup usually went like this. Let's say you wanted to build a client portal where your customers could check their project status without emailing you every Monday. Or a booking tool that would finally replace the spreadsheet you'd been copy-pasting for two years. So you opened ChatGPT.

You typed something like "Build me a client portal where users can log in and see their project status." ChatGPT responded with an intimidating wall of HTML, CSS, and JavaScript. But it looked promising. You copied the code, pasted it into a file, and opened the file in your browser. The page loaded, sort of. The login form was sitting there, but clicking the button did nothing. So you went back to ChatGPT, described the problem, and got an apology plus more code. You pasted that in. Now the button worked, but it threw an error about a missing database connection.

You didn't actually know what a database connection was, but you described the error to ChatGPT and pasted whatever fix it sent back. A new error appeared, something about environment variables. You asked about it, got an explanation and a configuration file, and pasted that too. The original error went away, but the page just showed a blank screen. So you went back and described the blank screen, and ChatGPT gave you yet another version of the code. This one was completely different from what you'd been building. The login form had disappeared, the styling looked different, and you weren't sure ChatGPT was even trying to solve the same problem anymore.

Three hours later, you had forty-three messages in a ChatGPT thread, four different versions of code that contradicted each other, and nothing that worked. You closed the tab.

And this wasn't just happening to non-technical people trying to do technical things. A developer on DEV Community described the same cycle: getting stuck on a bug, opening ChatGPT, pasting the error, getting a wall of code back, copying it into the project, hitting a different error, pasting that one too, and repeating the process for months. He eventually realized he was "using AI like a glorified Stack Overflow" and wasting more time than he was saving.

The cycle wasn't a prompting problem. You could write the most detailed, perfectly structured prompt in the world and still hit the same wall. When you paste a function into ChatGPT, the AI sees only text. It can't see what's in your project folder, has no idea what other files exist, and doesn't know what version of anything is installed on your machine or what you changed ten minutes ago. It's like texting a contractor a few photos of your kitchen and asking for a renovation quote. They'll do their best, but they're guessing at the plumbing, the wiring, the load-bearing walls, and everything else that doesn't show up in the photo. You're asking it to fix a problem it can't see.

The real issue was architectural. A chatbot generates text. To build a working application, you needed something that could read the entire project, understand how the pieces connect, make changes to the right files, run the code, see the errors, and fix them. Copy-paste coding can't do any of that because the AI never has the full picture. In the AI space, this "full picture" is called context and you'll hear about it a lot throughout this book.

If you experienced any of this, you probably already sensed that the problem was structural even if you couldn't have explained it that way at the time.

WHY CLAUDE CODE IS CHANGING THE AI GAME

Claude Code is built differently. It's more like the contractor who walks through your whole house in person, measures the rooms, checks the systems behind the walls, and comes back with a complete plan based on what's actually there. In other words, it can see your entire project.

When you start a Claude Code session inside a project folder, Claude reads everything that's in there. That includes every file in the folder. Your code, your design files, the settings that tell everything how to run, the folder structure itself, even the notes file most projects have at the top. Instead of looking at a snippet you pasted in, Claude has the whole picture of how your project is organized and how the pieces connect.

This changes what becomes possible because, instead of copying code back and forth between a chat window and your project, you describe what you want the way you'd describe a meal to a server or a vision to an interior designer. Claude Code does the rest. It plans an approach that spans multiple files, writes the code, runs it, and checks for errors. When something breaks, Claude reads the error, figures out what went wrong, and fixes it. Your job stays at the level of reviewing what came back to see if Claude actually built the thing you wanted it to build.

Go back to the client portal example. Instead of "here's some code, go paste it somewhere," Claude Code would read your existing project files, create the login page in the right location, set up a database connection, write the authentication logic, run the whole thing to confirm it works, and hand you back a working feature. If something broke along the way, Claude would see the error in context, figure out which file caused it, and fix it without you ever copying a single line of code.

WHAT YOU CAN BUILD WITH CLAUDE CODE

The hackathon winners built ambitious things in one week, including medical tools, construction permit workflows, and generative music systems. Your first build can stay much smaller than any of those. What matters is that it solves a real problem in your actual life or work.

A few categories of projects work especially well as starting points. None of them need a team, and most can be built in a single afternoon.

One category is tools for the work you're already doing every week. Examples include a proposal generator for your consulting practice, an intake form that replaces the email back-and-forth with prospects, or a dashboard for your business or life that gives you all the important info you want to see in one view. You know these problems better than any developer would, because you deal with them every day. You've probably been patching the gap with a mix of spreadsheets, email threads, and paid software that does 80% of what you want and charges you monthly for the privilege. Claude Code lets you build the version that does exactly what you need.

Another category is automation for the stuff you already hate doing. I built one for my own business finances. Every month, I used to spend an hour reconciling transactions, categorizing expenses, and pulling together a quick report on where the money went. Claude handles all of it now from a single command. The same approach works for personal finances, or really for any repetitive task you do every week and have been meaning to set up a better system for. As long as you can describe the steps to a person, Claude can build a version that runs the whole thing itself.

Websites and landing pages are another fit, especially when you want to validate an idea before sinking serious time into building the full thing. Lakshmi Narasimhan, a developer who documented his process publicly, used Claude Code to build a managed hosting landing page with a waitlist signup in under an hour. He spent twelve dollars on a domain and nothing else. He treated the page as a hypothesis test. If fifty people signed up, he'd build the full product. The same pattern works for any idea you're not sure about. You build the page, put it in front of your audience, and let the response tell you whether the thing is worth building.

Before we dive in, you're probably wondering who ModernMind Publications is and what makes us qualified to write this book.

ModernMind Publications is the publishing arm of ModernMind AI, run by two women, Lauren and Mary. Both of us come from project management backgrounds in the software space, which means we spent years working alongside developers, watching software get built, and learning the difference between what's hard and what just looks hard. Neither of us is a developer. If you asked either of us to sit down and write the code for a working app from scratch, the answer is no.

What we do is build with Claude Code. ModernMind AI vibe-codes custom products for small businesses, sets up Cowork automation and workspace buildouts, and runs AI consulting for owner-operators who want AI integrated into how their business actually runs. Everything in this book is the same approach we use with paying clients, distilled into a beginner path.

I'm Lauren, and I write the Field Guide series — so when I say "I" throughout the chapters, that's me. Any stories about building tools for our own operation are mine.

Two project managers who learned to build with AI, teaching you how to do the same.

This book walks you through building one real tool from scratch in Chapter 4, and you'll use that tool as your example throughout the rest of the book. The larger goal is bigger than any one finished project though. By the last chapter, you'll have a framework for thinking about *what* you can build and how to approach building it, so you can spot the next opportunity in your own life or business without needing a book to walk you through it. The hackathon winners weren't standout engineers. They lived close to their problems and had the right tools to build the fix. You have the same tools they did.

2

Setting Up Your Workspace

If you've been online looking for Claude Code tutorials, every YouTube walkthrough, every blog post seems to show the same thing: a black window full of white text and someone typing commands that look like they belong in a hacker movie. You used to have to work from the terminal on your computer to use Claude Code.

But in early 2026, Anthropic released a Desktop app for Claude Code. It's a regular application you download and install on your computer. You'll see a chat panel and clickable buttons. You won't see a black window that makes you feel like one wrong keystroke could set your computer on fire.

The Desktop app isn't a simplified version. Anthropic's own docs call it a friendlier way to access the same engine. Same capabilities, same results. The only difference is how you interact with it. If you've ever used Slack or Notion or any app with a sidebar and a main panel, you will likely feel a sense of familiarity using the desktop version of Claude Code.

This book uses the Desktop app for everything. If you'd rather use the terminal, the install instructions and prerequisites are in Appendix A.

WHAT YOU NEED BEFORE YOU INSTALL

The installation itself takes less than five minutes. First, you need a Claude account with a Pro subscription at $20 per month. That gets you what you need for everything in this book. It includes the latest versions of Anthropic's flagship models, Opus and Sonnet, extended thinking for complex problems, and a context window up to one million tokens depending on the model (more on that in Chapter 5). Everything you need to build real software is in there.

Pro is one of four tiers. Free only gives you chat access, not Claude Code itself. Max 5x at $100 a month and Max 20x at $200 a month exist for people who code heavily every day and burn through the Pro quota. I use the $200 Max 20x plan because I use Claude Code daily to build and run systems for my businesses.

If you're just starting out, Pro is perfect. Max doesn't give you better AI, it just gives you more usage before you hit a limit. We'll cover how usage relates to cost, and how to stay in control of it, in Chapter 8. For now, my recommendation is to pick Pro and move on.

Second, your computer needs to meet some basic requirements: macOS 13 or later, Windows 10 (build 1809 or newer) or Windows 11, or Linux with Ubuntu 20.04 and up. You need at least 4GB of RAM and an internet connection. If your computer was made in the last five years, you almost certainly qualify.

On a Mac, click the Apple menu in the top-left corner, then click "About This Mac." The version number is right there. On Windows, open Settings, go to System, then About. You'll see your Windows version and build number. If either of those shows something older than what I just listed, you'll need to update your operating system before continuing.

INSTALLING AND LAUNCHING FOR THE FIRST TIME

Everything should be in place at this point. You have a subscription. Your system meets the requirements. Let's get Claude Code running.

To download the Desktop app, go to claude.ai `/download` in your browser. You'll see download options for macOS and Windows. Click the one that matches your computer, run the installer, and open the app once it finishes.

When you open the app for the first time, it'll ask you to sign in. Your browser will open to a Claude login page. Sign in with the same account you used to set up your Pro subscription. Once you authenticate, Claude Code stores a security token on your computer so you don't have to sign in again. This happens once. Before 2026, setup meant copying API tokens by hand, pasting them into config files, and troubleshooting when they expired. The browser sign-in replaced all of that.

If the app opens and you see a chat interface with a text input area, you're done. Claude Code is installed and running.

WHAT YOU'RE LOOKING AT

You should now be in what's called your workspace. In the top left corner is a bar that shows Chat, Cowork and Code. Chat handles conversations, Cowork connects your tools and runs workflows, and Code builds software. We're working in Code throughout this book, so click that tab.

This takes you to a new view where, in the center, there's a chat panel where you type prompts and read Claude's responses. Right above the chat bar you'll see a control labeled Project folder. That's where you tell Claude Code which folder to work in. Whatever folder you select becomes Claude's world. It can see everything inside it, and that's where it'll create or change files. You can switch folders anytime by clicking the Project folder control again and picking a different one.

Below the chat bar on the right is a model selector for choosing which version of Claude powers the session.

I always do a small throwaway test like this when I'm setting up a new tool, just to get a feel for the process. Create a new folder somewhere on your computer (Desktop is fine, call it claude-test or anything you want). In Claude Code, click the Project folder control above the chat input and select the folder you just made. When the trust prompt comes up, say yes.

Now type "hi" into the chat bar and send it. Once you start a chat, an icon will appear in the top right corner that says "Views" when you hover over it. Clicking it presents a dropdown. That's where the rest of the app's capabilities live. For now, click it and select Files. A panel opens on the right showing the contents of the folder you selected with the Project folder control. I find this view more reassuring than anything else in the app, because I can see exactly what Claude is looking at before it does anything.

The other things tucked into Views (Diffs, Previews, Plan Mode, and a built-in Terminal) become relevant in later chapters. You don't need them yet.

Now type this prompt into the chat panel:

```
Create three folders in this directory called projects,
clients, and ideas. Inside each one, add a README.md
file with a sentence or two explaining what goes in
there.
```

Send it.

Claude will think for a moment, then come back with a list of files it wants to create. Read what it's proposing. If it looks right, approve. The folders and files appear in your folder on your computer and also in the Files panel in the sidebar on the right.

THE CORE LOOP

You type a prompt. Something like "read the files in this folder and tell me what they do" or "create a simple webpage that shows my business hours." Claude reads the files in your project folder, thinks about your request, and responds. If the response involves changing or creating files, Claude shows you the proposed changes first. You read what it wants to do. If it looks right, you approve, and Claude applies the changes to your actual files. If something looks off, you tell Claude what to adjust, and it proposes a revision.

The approval step is the safety net. Claude doesn't silently change your files. Every modification gets proposed first, and nothing happens until you say yes. For someone who's new to all of this, knowing that nothing changes without your explicit permission can provide peace of mind while you're still trying to figure out what you're doing.

You'll get permission prompts beyond just file changes. Claude will also ask before running commands on your computer or using a connected tool like your calendar or email. The decision rule stays the same. Approve if the action matches what you asked for. Otherwise, say no and ask Claude what's going on. Chapter 7 covers the full permission system, including the modes that let Claude work with more autonomy once you're comfortable with the basics. Every chapter that follows builds on this same loop. The projects get more complex and the prompts get more specific. But the interaction pattern stays exactly the same.

3

Writing Prompts that Lead to Quality Output

Nathan Onn, a developer who documented his debugging process in detail, named the central problem of building with AI in one sentence. "Claude almost never fails because it lacks intelligence. It fails because it lacks visibility."

That visibility problem is what this chapter is about. You're going to build your first real tool in the next chapter. The biggest difference between the build that takes an afternoon and the build that frustrates you for three isn't the model. It's how much of the right information Claude has when it starts working. So this chapter teaches the skill before you need it.

Think about what Claude can actually see when you type a prompt. It sees your words. Not your screen, not the runtime behavior of your code, not the error message that just flashed in the browser, not the thing in your head that you forgot to mention. Claude writes code based on your words, hands the code over, and waits. If something breaks, Claude doesn't know unless you tell it, and unless you tell it specifically.

"Make the form better" forces Claude to guess what "better" means. It could mean fewer fields, faster load time, a different layout, cleaner validation, better mobile behavior. Claude picks something, and it's probably wrong, because "better" isn't information. "Add validation to the email field that rejects addresses without an @ sign and shows a red error message below the field" builds exactly that. The model didn't get smarter between those two prompts, it had more to work with.

THE TWO LAYERS OF CONTEXT

Context has two layers, and you need both of them to get a clean result on the first pass.

Background is who you are, what you're building, and why. It's the frame Claude uses when making decisions you didn't explicitly specify. "I run a freelance web design business, and I want an intake form on my website so I can stop chasing prospects over email for information I should have collected the first time they reached out." That's background. Claude now knows the form is client-facing, the tone should match a professional services business, the reader is a potential customer, and the job is gathering enough information to respond intelligently when that prospect turns into a lead. You didn't tell Claude any of those things literally. Claude inferred all of them from two sentences.

Background looks different depending on who you are. A solopreneur running a fitness coaching business gives different background than a founder building a SaaS prototype than a freelancer scaling client services. The common thread is the frame, not the specifics. Who is this for, what outcome are you trying to reach, and what existing reality does the tool need to fit inside. Two to four sentences are usually enough. You don't need to write a full biography.

Requirements are what the thing specifically needs to do. What goes in, what comes out, what happens in between. "The form collects the client's name, email, phone, project type, and budget. The submit button shouldn't work until the required fields are filled in. When someone submits, I want to see their info somewhere I can check later, and I want them to see a message letting them know I got it." No mention of HTML, JavaScript, databases, or single-page applications. You described what the thing does, not how it's built.

Here's the move that makes this work for non-technical people. You don't have to know every requirement up front. You write down what you know and tell Claude to ask you questions about everything you didn't cover. Add a single line to the end of your request.

```
Before you start building, ask me any questions you
need to understand what I'm looking for.
```

What comes back is a list of questions that force you to make decisions you didn't know you had to make. Should submissions be saved to a file on your computer, emailed to you, or sent to a spreadsheet? Should the form clear itself after someone submits, or show a confirmation page? If a prospect enters an invalid phone number, should the form reject it or accept it and flag it for you to review later? Each question has a plain-language answer. You didn't need to know the technical terms for any of these choices. You just had to recognize which option matched what you wanted.

Claude is extremely good at translating plain-language descriptions into technical decisions when you let it. What it can't do is ask you things you haven't invited it to ask about. That one line at the end of your request is the invitation. Without it, Claude guesses. With it, Claude asks.

There's a move I use constantly that takes this one level deeper. Once Claude has asked its questions and you've answered them, ask Claude to write the prompt for you.

```
Based on everything we've discussed, write the prompt I
should paste into a fresh session to get exactly this
built.
```

Claude takes the conversation you just had and compresses it into a single structured request that captures every decision and constraint you worked out together. You open a new session, paste that prompt, and Claude starts from a clean slate with the full picture already in place.

The first session does the thinking. The second one does the building. It sounds like extra work, but it's faster than trying to build inside a session whose context is already half-polluted by the back-and-forth of figuring out what you want. I use this pattern almost every time I'm building something I haven't fully thought through yet, and it's one of the biggest unlocks I've found for getting clean output on the first real attempt.

CONTEXT IS A SKILL YOU DEVELOP BY FAILING AT IT

Nobody's good at this on their first try. Not me, not the developers in the case studies, not the non-technical founders who are shipping tools they built themselves. You get good at providing context by repeatedly failing to provide enough of it and noticing what broke.

Here's what failing looks like. You're a few messages into building an order form for your website. The form works and submissions are saving. You ask Claude to add an email notification so you get a message whenever someone submits. Claude builds the feature using a library that assumes your site runs on a certain kind of hosting. Your site runs on the basic shared hosting plan that came with your domain name, which doesn't support that library. You didn't know to specify your hosting situation because you didn't know there were different kinds of web hosting.

The emails don't send. You tell Claude the emails aren't sending. Claude tries a different approach. That one doesn't work either, because the underlying issue is the same. You go back and forth for forty minutes, the context window fills up with failed attempts, and by the end of it the session is too polluted to rescue. You have to `/clear`, start over, and explain your hosting setup at the top of the new session. Problem solved in ten minutes once Claude knew what it was working with.

The detail you didn't provide, where your website is hosted, cost you forty minutes and a dead session. Not because Claude did anything wrong. Claude made a reasonable assumption based on the information it had, and that assumption didn't match your reality. You couldn't have known to specify your hosting environment because you didn't know it mattered. Next time you will.

Every time Claude builds something that misses, there's usually a piece of context you could have handed over that would have prevented it. The missing piece might be about your technology environment, like which hosting plan you use, which database the project writes to, or which browsers your customers are on. Sometimes it's about your business, like who the reader is, how formal the tone should be, or what a reasonable price range looks like for your market. Other times it's about the project itself, like what existing pieces Claude shouldn't touch or what style the rest of the app follows.

You don't need to figure out every possible piece of context before you start. You need to start, notice when Claude misses, figure out what it didn't know, and hand that over at the top of your next attempt.

Practice it deliberately and you'll be dramatically better in a month. Skip the practice and you'll still be typing "make the form better" six months from now, wondering why Claude keeps missing.

This is also why I don't want you to chase a perfect first prompt. Perfect prompts are a myth, and trying to write one keeps people stuck at the starting line. The real skill is iterating fast. Write what you have, let Claude ask questions, build something, notice what's missing, and carry the lesson into the next session. That loop gets tighter every time you run it.

THREE MOVES THAT MAKE CONTEXT EASIER TO DELIVER

Claude Code has three features that make context delivery easier, especially for non-technical people. Each one replaces a kind of context you would otherwise have to put into words.

File references with @. Any time you mention a file with the @ symbol, Claude reads that file before it starts working. "Follow the layout pattern in @components/`Header.jsx` for the new component" tells Claude exactly where to look instead of asking it to hunt through your project. You're not asking Claude to guess at your codebase. You're pointing it at the exact file you want it to mirror or modify.

If you're not sure what files exist in your project, ask Claude to list them first. "List the main files in this project and give me a one-sentence description of each." You get back a map of your project, and the names become reference points for your next request. The @ syntax is available in both the Desktop app and the terminal, and it works the same way in either one.

Screenshots. Nomiki Petrolla is a non-technical founder building Theanna, a platform connecting over 300 women founders. She doesn't use technical terminology. Her entire workflow is plain English descriptions and screenshots. She takes a picture of what she wants something to look like, drops it into Claude Code, and describes the changes she wants. Her Challenge Prompts feature would have taken at least two weeks through the traditional process with Figma mockups, requirements documents, and engineering handoffs. With Claude Code, she shipped it in under two days. She never wrote a structured prompt template or learned technical vocabulary to do it.

The screenshot is the context. Claude can see what you see, which means it knows what you mean. This matters most when you're describing visual things (layout, spacing, colors, the feel of something) where words run out faster than you'd expect. "Make it look like this" paired with an image does more than a paragraph of adjectives.

The same move works with text. If you want Claude to write in your voice, paste two or three things you've already written and say "match this tone and structure." If you want new code to follow a pattern already in your project, paste the existing file and tell Claude to mirror it. Examples do work that adjectives can't. Your voice, the tone your audience responds to, the level of polish your work needs, all of these are faster to show than to describe. When you have a good example, hand it over instead of reaching for more words.

Plan Mode. Before you ask Claude to build anything with more than one or two moving parts, press `Shift+Tab` twice. Claude switches from writing code to proposing an approach. It looks at your project, thinks through what needs to happen, and lays out a plan for you to review. You read the plan the way you'd read a contractor's estimate before work starts. If something looks off, you say so before any code gets written.

Chapter 7 goes deep on Plan Mode. For now, just know it exists and use it when a request feels bigger than you can fully describe. The plan Claude produces often surfaces questions and assumptions you didn't know you had. Answering those questions at the planning stage is dramatically cheaper than discovering them after Claude has rewritten three files.

Plan Mode gets its own chapter because it matters that much. So does `CLAUDE.md` in Chapter 6, a file Claude reads automatically at the start of every session so you never retype your background. Keep building the underlying skill here. Those chapters show you how to deliver context without typing it from scratch.

FIXING A BAD FIRST PASS

You gave background, described the requirements, and let Claude ask questions. Something still came back wrong. What now?

This is where most people fall into a correction loop. They type "that's wrong, fix it" or "the button doesn't work" and hope Claude figures out what they meant. It won't. Claude still can't see what happened when the code ran. You have to be its eyes.

Four patterns cover almost every correction you'll run into.

Show, don't argue. When something breaks and there's an error message, paste the exact text of the error. Don't describe it, don't paraphrase it, don't reword it to sound more technical. Copy the error and hand it over.

Nathan Onn was debugging a React component where a search filter broke when someone also used the category filter. Instead of telling Claude "the filter is broken," he added a few logging statements that watched what was happening as the component ran, then pasted those logs directly into Claude. The logs revealed that two pieces of code were running in the wrong order, and the second one was overwriting the first. Claude's fix was surgical because Claude could see the runtime behavior, not just hear a description of the symptom.

Showing the logs usually means opening your browser's developer tools (right-click the page, click Inspect, then click the Console tab) and copying whatever red error message appears when the thing breaks. You don't need to understand what the error says. You just need to hand it to Claude.

Expected versus actual. When the output is wrong but there's no error message, describe the gap between what you expected and what you got. "The form should reject submissions when the email field is empty. It's accepting empty submissions." One sentence, three pieces of information. The intended behavior, the actual behavior, and the location of the mismatch. Compare that to "the form validation isn't working," which tells Claude nothing about what's broken or where to look.

Diagnose before fixing. When you don't know what's wrong, don't ask Claude to fix it. Ask Claude to diagnose and report back without changing anything.

```
Something's off with the form. Investigate what's
happening and tell me what you find, but don't make any
changes yet.
```

Read what Claude reports. Then decide whether to fix it in the current session or start fresh with a more targeted request. Separating diagnosis from fixing keeps Claude from making speculative changes that introduce new problems while chasing the old one.

Know when to stop. After two failed corrections on the same issue, stop. Don't try a third time. The context is already polluted with failed attempts and conflicting instructions, and Claude isn't reading your original request anymore. It's trying to reconcile contradictions you didn't mean to create. Type `/clear` to reset the conversation, then write a new request from scratch. Incorporate what you learned from the failed attempts. What worked, what broke, and what do you need to specify differently this time?

This is Anthropic's official recommendation, not just community advice. It feels counterintuitive because it looks like starting over. It's actually the fastest path to a working result, because you're giving Claude a clean context with better information instead of asking it to untangle a mess.

Three session controls make all of this smoother. Esc stops Claude mid-action if you see it heading in the wrong direction. Esc twice or `/rewind` restores the previous state if Claude made changes you want to undo. And `/clear` resets the entire conversation when it's time for a fresh start. These exist because the correction problem is real and common enough that Anthropic built escape hatches into the product.

`/rewind` is worth highlighting. It's an undo button for your entire project. If Claude changed five files and the result is worse than what you started with, `/rewind` puts every file back exactly where it was before Claude touched them. One command, everything restored. You don't have to remember what changed or figure out how to reverse it manually. This alone takes most of the fear out of letting Claude try something ambitious.

WHERE YOU GO FROM HERE

When you build the tool in the next chapter, watch for the correction loop. You'll probably spot at least one place where a `/clear` and a better-prepared starting message would have saved you ten or fifteen minutes. Everyone does it. The difference between the people who stay stuck there and the people who ship real tools is noticing the pattern and responding differently the next time.

What you learned in this chapter is the skill. The chapters that follow give you the tools that make the skill easier to apply. Chapter 5 covers how Claude Code holds information in its head, which is what makes the "know when to stop" rule work. `CLAUDE.md`, introduced in Chapter 6, is the file that stores background so you never have to type it into a prompt twice. Chapter 7 walks through Plan Mode, the deliberation step that catches assumption mismatches before Claude writes a line of code. And Chapter 8 shows what sloppy context actually costs you in dollars, not just minutes. Every one of them is about the same thing as this chapter. How much of the right information is Claude working with, and how do you make sure the answer is enough.

4

Creating Your First Tool

Vague ideas are the real killer of first builds. A "client portal" is a six-month project that never finishes because the scope is too ambiguous. Now compare that to "a form that collects five fields and saves them to a file," which is a two-hour project you can ship today.

A good first build is something you'll actually use yourself once it's done. That motivation is what makes you bother to finish it. You can describe what the thing does in two sentences without leaving anything out. And it shouldn't need user accounts or payment processing, because those add weeks of work that teach you nothing about the parts you actually want to learn.

We're building a client intake form. It collects five fields from whoever fills it out: name, email, phone number, project type, and budget. The submission gets saved to a file on your computer, and the person who filled it out sees a confirmation message.

If you run any kind of service business, you've already collected client information the slow way. It was probably a back-and-forth email thread, or a DM where the client answered three of your five questions and you had to chase the rest. A form standardizes that part of the work and stops you from doing the chasing.

The intake form covers two things you'll need everywhere. One is handling form input. The other is saving data. It doesn't need a backend server, which means you don't have to learn what one is yet. You'll open the finished tool in your browser the same way you open any website. Later chapters add features to this same form. We'll get to debugging, then security, then deployment. For now we just need it to work.

A landing page takes most people thirty minutes with Claude Code. A form that saves data, like the one we're building, takes two to four hours once you include the back-and-forth. Anything with a real backend and a database is closer to a weekend of work. The intake form sits in the middle on purpose. It teaches you every step of the build cycle without burning a weekend.

Project type	Example	Estimated build time
Simple static tool	Landing page, single-function calculator	10–30 minutes
Form-based tool with data storage	Client intake form (this chapter's build)	2–4 hours, including iteration
Tool with databases & user accounts	Small dashboard, lightweight admin panel	1–2 days
Multi-feature app	Scheduler, CRM module, full client portal	3–5 days, phased

Table 4.1: Project complexity and estimated build times

An idea to a working tool, in one afternoon. That's the test for whether the project is the right size.

OTHER PROJECTS THAT WORK FOR A FIRST BUILD

The intake form is the example I'm walking through throughout the entire book. If you have your own idea you'd rather build, use that one instead. The steps work the same way.

Some other projects that fit the criteria above:

1. **Client intake form (what this chapter walks through).** It collects name, email, phone, project type, and budget, and saves each submission so you can follow up.

2. Quick proposal generator. You fill in the service type, the client's name, and the scope of work, and it produces a formatted proposal you can copy into an email.

3. Photo-to-funny generator. You upload a photo and Claude turns it into something playful. Maybe a dramatic movie-trailer narration of what's happening, a caption written like a medieval history book, or a thought bubble for your pet. Pick whichever direction makes you laugh.

4. Nutrition tracker with feedback. You log what you ate each day, and over time the app spots patterns and tells you what to tweak. Low protein on Tuesdays. More sugar after bad sleep. It's like a personal coach that watches your habits and points out what to focus on.

5. Your own idea. If you see a problem in the world that nobody has solved the right way, build the first version of it. You can start with a landing page and a waitlist to test whether anyone cares, or go straight into building the thing itself. The best entrepreneurial builds come from a problem you understand better than anyone else does.

Just remember to try to keep it simple for now. Halfway through, you'll get an idea. Maybe a dropdown for project urgency, an email notification when a form comes in, or a way to export everything to a spreadsheet. Don't add it yet. Each new feature has its own bugs, and you've barely tested the original ones. So finish the form first and use it for a week. The features you actually need will get clearer over that week, and the ones you only thought you needed will quietly drop off the list.

TELLING CLAUDE WHAT TO BUILD

Chapter 3 taught you the skill of writing a brief. This section applies that skill to a specific project.

"Build me a form" leaves Claude guessing about every detail you didn't say out loud. Which fields belong on it, what happens on submit, where the data lives, what the thing looks like. Some guesses will be right and some will be inventions you'd never have asked for. Claude isn't slow because it's bad at building forms. It's slow because it doesn't know your business yet, so it has to fill in the gaps you didn't.

Wrong guesses compound. When Claude makes one and you don't catch it on the first pass, that guess gets baked into the project. The next session, Claude opens the code and treats the wrong design as your design. By the third session, you're arguing against decisions nobody actually made on purpose.

A good prompt tells Claude where you're starting from, what you want it to do, and what counts as done. Chapter 3 covered the first two of those. The "done" part is the one specific to building tools, and it's also the part most beginners skip. Their projects grow without ever finishing.

Type this prompt exactly as written, or adjust the details to match your own business:

```
I want a client intake form I can put on my website. It should collect the client's name, email, phone, the type of project they want (website, app, or other), and their budget. All the fields are required except budget. The submit button should only work when all the required fields are filled in. When someone submits the form, save what they entered somewhere I can check later, and show them a message letting them know I got it. Use clean, professional styling that would look right on a freelance business website. When you're done, I should be able to open the form in my browser, fill it out, submit it, see the confirmation, and verify the submission was saved. Before you start building, ask me any questions you need to understand what I'm looking for.
```

The prompt runs about 135 words and took a couple of minutes to write. It reads the way you'd describe a project to a colleague in the break room. Claude handles the technical translation, which means you don't have to say anything about HTML or JavaScript or whatever framework will end up being used. The clearer your description is, the smaller Claude's room to guess.

When you paste this prompt into Claude Code, the first thing that should happen is Claude asking you questions. Where should submissions be saved? What should the confirmation message say?

Should the form clear itself after submission? Every answer closes a decision Claude would otherwise have made on its own.

TIME TO BUILD: THE FULL WALKTHROUGH

Open Claude Code the way you set it up in Chapter 2. First, create a new folder on your computer called my-intake-form. Put it wherever you keep work projects, or drop it on your Desktop if you don't have a spot yet. Then in the Claude Code app, find the Project folder control above the chat input, click it, and select the folder you just made. The app loads that folder as your workspace with an empty chat ready for your first prompt.

Paste the entire prompt from the previous section.

Watch what Claude does next. It doesn't write code. It reads your prompt and pauses. Then one of two things will happen. Claude will ask a question about something you didn't specify, which is what the last line of the prompt asked for. Or, if the prompt is complete enough on its own, Claude will start building. Either response is good. The bad one is silent code generation off a vague brief, which is what happens when you type "build me a form" and walk away.

Once you approve the plan or answer Claude's questions, the building starts. Files appear in your project folder. For an intake form, you'll usually get three files: an `index.html` for the page itself, a `styles.css` for the look, and a `script.js` for what happens when someone clicks Submit.

Claude writes the code, tests it, and shows you the results. Sometimes you'll see it catch its own mistakes along the way, like fixing a validation rule that wasn't quite right or adjusting the styling. That self-correction is normal. It's one of the reasons Claude Code works differently from copying and pasting code from a chatbot. Claude can see the whole project, run the code, notice when something breaks, and fix it without you needing to step in.

When Claude finishes, it'll tell you. Open your file browser, navigate to the my-intake-form folder, find `index.html`, and double-click it. Your browser opens. There's your form. Five fields and a submit button, sitting on a clean page that didn't exist twenty minutes ago.

Test it. Type a name, an email address, a phone number. Pick a project type from the dropdown. Leave the budget field empty since it's optional. Click Submit. You should see a confirmation message, and the form should clear itself, ready for the next entry.

Now check whether the data actually saved. Look in your project folder for a JSON file, probably called something like `submissions.json` or `data.json`. Open it in any text editor (Notepad on Windows, TextEdit on Mac, or whatever you use to open plain text files). You should see the submission you just entered, with name, email, phone, and project type all stored as structured data.

Take a second with that. You described what you wanted in plain English. Claude built it. You opened it in a browser and it worked. The thing that used to take a freelance contract or a learn-to-code year just took an afternoon.

But it probably isn't perfect. The form might look a little plain. The confirmation message could disappear too fast, or not disappear at all. Maybe you tried submitting without filling in a required field and the error message wasn't clear. The layout might look weird on your phone. These small imperfections are normal.

William Roberts had zero programming experience when he started with Claude Code. Within six months, he shipped five production applications totaling over 300,000 lines of code. His first app took a few days. That whole trajectory started where you're standing right now, with a first build, a few imperfections, and the realization that the process actually works.

THE ITERATION LOOP: WHERE THE REAL LEARNING HAPPENS

You're looking at your intake form and something isn't right. The submit button might be active before all the required fields are filled. The confirmation message could be covering the form instead of appearing below it. Or the project type dropdown still says "Option 1, Option 2, Option 3" when it should say "Website, App, Other."

Whatever it is, you can see it. That noticing is where building with AI actually starts. You don't need to know how to fix it yet, just to know it isn't right.

Go back to Claude Code and describe what you found. Be specific. "The submit button should be grayed out until all required fields are filled" is a fix Claude can act on. "The button doesn't work right" is a guess Claude has to interpret, and it might interpret it wrong.

Claude reads your feedback, looks at the existing code, makes the change, and tests it. You refresh the browser and check. Better? Good. Not quite? Tell Claude what's still off. Each cycle gets the form closer to what you actually want.

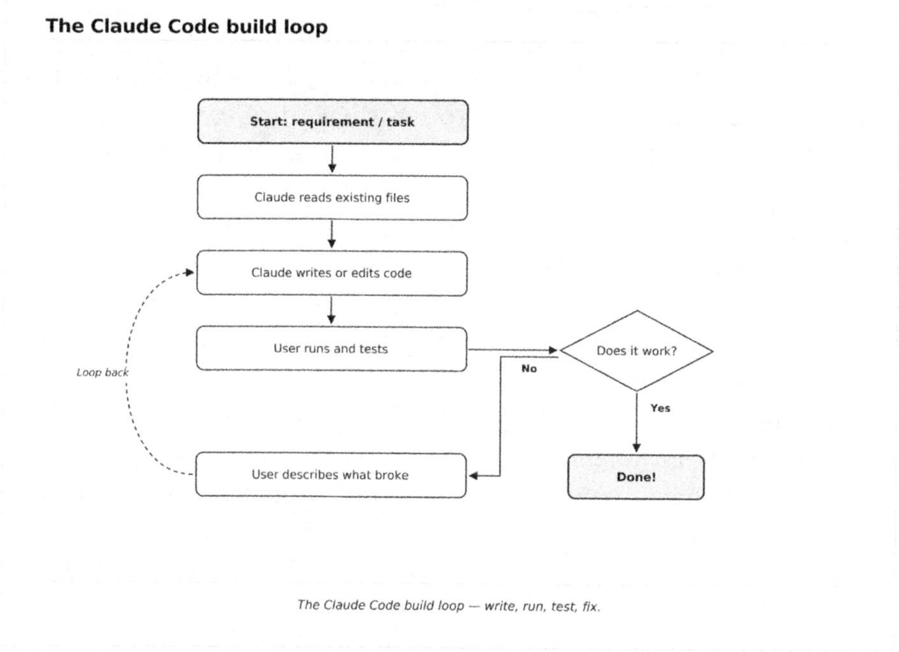

Figure 4.1: The Claude Code build loop — write, run, test, fix.

Each round takes a minute or two. By the third or fourth round, the form does what your prompt described. After a few more rounds, it does things that weren't even in the brief, things you only noticed once you started using it. That's iteration working as it should.

Most projects go sideways at exactly this point. The iteration loop feels productive, and it is. But it's also where scope creep sneaks in. You fix the validation, then think "what if I added an email notification when someone submits?" You fix the styling, then think "what if there were a dashboard showing all submissions?" Each feature sounds small on its own. But each one adds complexity, and complexity is what kills most projects somewhere past the halfway point.

Most vibe-coded projects (the community's term for software built by describing what you want to an AI) don't die because the code is bad. They die because the builder kept adding features during the iteration phase, the project got more complicated than the original brief, and the whole thing collapsed under its own weight. The builder walks away thinking the tool didn't work. In reality, the tool worked fine. The scope didn't.

Finish the intake form as described in your brief. Test it. Use it for actual client inquiries if you can. Live with it for a day or two. Then, in a fresh Claude Code session, add one feature. Then another. Working tools become working systems through small, deliberate additions. A week of careful additions will get you further than any marathon session.

Brant Choate, CTO of Remi, saw this pattern when he sent a Claude Code tutorial to his non-technical employees. They started building tools they actually used at work. The pattern was the same every time: a small, well-scoped first project, an iteration loop to get it right, and gradual expansion from there.

But the iteration loop has a failure mode worth catching before it catches you. A developer named Tejas asked Claude Code for what should have been a simple component and wrote up what happened on DEV Community. The first version came back with one thing he wanted changed. Easy correction. He asked Claude to fix it. The fix worked, and broke a different feature. He asked Claude to fix that one without losing the first. The second fix worked, and broke a third feature. By round four, Claude had hallucinated a library that doesn't exist and reverted the entire component back to the version Tejas had started with. Thirty minutes and four corrections produced no progress at all.

Tejas named this the Circle of Doom. Each correction introduces a new constraint that conflicts with an earlier one. Claude overcorrects, fixates on whatever you said most recently, and loses sight of the original goal. By the time you've invested enough time that writing the code yourself feels like admitting defeat, the first version was usually closest to what you wanted, buried seventeen messages deep.

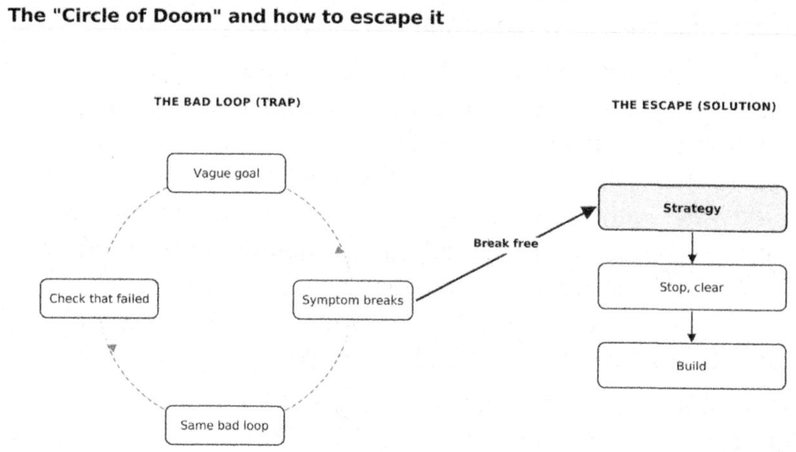

The Circle of Doom — and how to escape it.

Figure 4.2: The Circle of Doom and how to escape it.

Type `/clear`, take a minute, and start a fresh session with a prompt that includes everything you've learned about what went wrong. Anthropic's own documentation recommends this after two failed corrections on the same issue, because by that point the conversation context is fighting against you. The chapters ahead show you how to set up sessions so the Circle of Doom hits you less often. For now, the move is just to notice the loop and not push through it.

WHAT YOU JUST LEARNED (WITHOUT REALIZING IT)

You might be wondering whether you actually learned anything. You followed instructions, pasted a brief, clicked around a form. It feels like Claude did the work and you watched. You practiced a few skills without anyone naming them out loud.

When Claude showed you the finished form, you looked at it and knew whether it was right. You didn't read the code. Nobody told you if it worked. You opened it, tested it, and your own judgment told you what was good and what needed fixing. That was evaluation, the foundation of everything that follows. Evaluation is what makes you the builder instead of just the prompter.

Then something was wrong, and you told Claude exactly what to fix. Not "this isn't right" but "the submit button should be disabled until all required fields are filled." Every time you described a problem in concrete terms, you were teaching yourself how to communicate about software without knowing the technical vocabulary. You said "the confirmation message should appear below the form, not on top of it" instead of learning CSS positioning rules. The result was the same. The path you took to get there was yours. That's specificity in correction, a skill that sharpens with every project.

At some point during the iteration loop, you stopped. The form worked. It did what the prompt described. Was it perfect? Probably not. But it was done enough to use, and you made that call. Knowing the difference between "not finished" and "finished enough" is what separates people who ship from people who tinker forever. That judgment is the third skill, and it's the hardest one to teach because nobody can make it for you.

So you walked out of an afternoon with three things: evaluation, specificity, and the judgment to call something done. Good builders use these skills constantly. Everyone else prompts and hopes. The rest of this book deepens each skill. You'll see why Claude gets worse as a session drags on, how to carry context across sessions, when Plan Mode stops bad builds before they start, and how to give Claude the project memory it needs.

Notice what you didn't learn today. You didn't learn to code. You didn't learn syntax or how JavaScript handles events. You learned to build. The two things look related but aren't. Coding is a technical skill that takes a few years to get good at. Building is a creative discipline. You decide what you want, you describe it, and you tell the tool when it's wrong. None of that takes years. It takes practice.

Look at what's on your computer. There's a folder called my-intake-form. Inside it are files you can open, use, and modify. Together they make a working tool that collects client information. You built it by describing what you wanted and refining what you got back. You didn't pick up a programming language, a framework, or a six-month bootcamp along the way. You described, you evaluated, you refined, and now you have something real.

5

How Claude Code Thinks

Gary Parker, a Senior QA Architect, was building Playwright automated tests for a user registration flow when Claude forgot what it had been doing an hour earlier. Parker had asked it to add handling for invalid email edge cases. Instead of building on the existing test suite they'd spent the session creating, Claude forgot the suite existed. It generated a brand new, incomplete test from scratch. Parker lost nearly an hour untangling the mess and fixing the inconsistencies the rewrite introduced.

What happened there has a name. Claude Code has a hard limit on how much it can hold in its head at once. Every prompt you type, every file it reads, every response it generates takes up space in something called the context window. When there's room, Claude tracks your decisions and builds on what came before. When the window fills up, the quality drops fast. You'll ask for a change and get a suggestion you already tried twenty minutes ago.

Another developer, Dibyanshu Kumar, hit the wall even harder during a complex refactor. Forty minutes into the session, deep in interconnected file changes, Claude started forgetting things. Then the session died outright. Kumar described the experience bluntly. Whatever the model was working on, whatever context it had built up was gone. There was no warning before the crash, just a dead session and forty minutes of reasoning lost.

Kumar and Parker hit the same wall from different angles. This isn't a bug, a skill gap, or something specific to Claude Code. It's how every large language model works, and once you understand the mechanism, you stop fighting it and start working with it. I hit the same wall early, and I hit it often. My instinct was to push through and try harder. The move that fixed it was the opposite. I started closing sessions sooner, scoping them tighter, and starting the next one fresh instead of picking up where I left off.

WORKING MEMORY, NOT STORAGE

Think about your desk. I mean your actual workspace, not a metaphor. You've got papers spread across it: the thing you're working on right now, some reference material you pulled up earlier, notes from a conversation, maybe a printout you forgot to put away. You can only think about what's physically in front of you. If you need something that's in a filing cabinet across the room, you have to go get it. And if your desk is so covered in paper that you can't find the thing you need, having a bigger desk only delays the problem.

Claude's context window works the same way. It's everything Claude can "see" at one moment, all the information Claude can reason about during a single conversation. The context window is working memory. It's the papers on your desk, not what's stored in the filing cabinet.

Claude measures desk space in tokens. A token is roughly four characters of text, or about three-quarters of a word. When you see a context window described as 200,000 tokens, that translates to about 150,000 words. You don't need to think in tokens day to day, but knowing the unit helps when you decide whether to keep going or start fresh.

System instructions take up a few thousand tokens at the start of every conversation. Project context files add more. Then there's the full conversation history, every message you've sent and every response Claude has given, in order. Every file Claude reads during the session goes on the desk, along with every command output, every result, and every error message. Each one takes up space, and nothing gets removed automatically.

Here's the part that bites you. Claude sends everything on the desk with every request. When you ask Claude a question on your fiftieth message, Claude doesn't just receive that question. It receives the system instructions, the project files, all forty-nine previous messages and responses, every file it read, and every command output, plus your new question. The entire desk gets picked up and handed over every time. That's why the desk fills faster than you'd expect. Most of what fills the desk isn't your work. It's Claude re-reading files it already read, re-running commands you already confirmed, and re-summarizing decisions you made an hour ago.

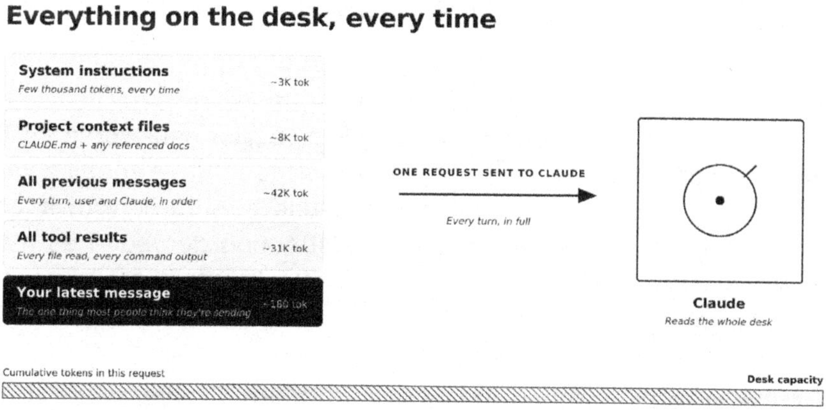

Figure 5.1: Everything Claude sends to the model with every single request (not just your latest message).

The numbers sound enormous at first. A 200,000-token context window holds roughly 150,000 words. That's a thick novel. Surely you won't fill that up in an afternoon.

You will. A complex build session can consume 50,000 to 100,000 tokens in a few hours of active work. Claude's responses are often longer than your messages, and they all stay on the desk. The internal thresholds, documented by security architect Ken Huang, fire when you're down to about 13,000 tokens remaining, with a warning at 20,000. Those are invisible tripwires. By the time you notice something feels off, you've already crossed them.

The models available in 2026 give you different desk sizes. Sonnet 4.5 and Haiku 4.5 have 200,000-token windows. Opus 4.6 and Sonnet 4.6 can use up to 1,000,000 tokens. More space helps, but it doesn't solve the problem.

In practice, I work in Sonnet most of the time and switch to Opus when the task needs nuance, like reviewing a complex piece of work or weighing tricky design choices. I never use Haiku for Claude Code. It's a smaller model, and for this kind of work the quality drop hurts more than the cost savings help.

WHY MORE SPACE DOESN'T FIX IT

The obvious response to everything you've read so far is to get a bigger context window. If 200,000 tokens fills up in a few hours, use a model with 1,000,000 tokens. Problem solved.

Except it isn't.

Chroma, a research lab focused on AI retrieval, tested 18 frontier language models and found that every one performed worse as the input got longer, even on tasks as simple as repeating a specific word from the text. The models didn't fail because they ran out of space. They failed because the quality of their attention degraded as more information accumulated. Chroma calls this "context rot," and the name fits: performance doesn't crash at a specific threshold. It decays gradually, like wood left in the rain.

There's a specific mechanism behind it. When Claude processes a long conversation, it pays the most attention to what's at the very beginning and the very end. Information buried in the middle receives less focus. Researchers Liu, Lin, Hewitt, Paranjape, Bevilacqua, Petroni, and Liang measured this directly. Accuracy dropped over 30% when the answer to a question appeared in the middle of the context rather than at the beginning or end. If you made a design decision twenty messages ago and it's surrounded by file contents, command outputs, and follow-up discussions, Claude may not give it the attention it deserves, even though it's technically still on the desk.

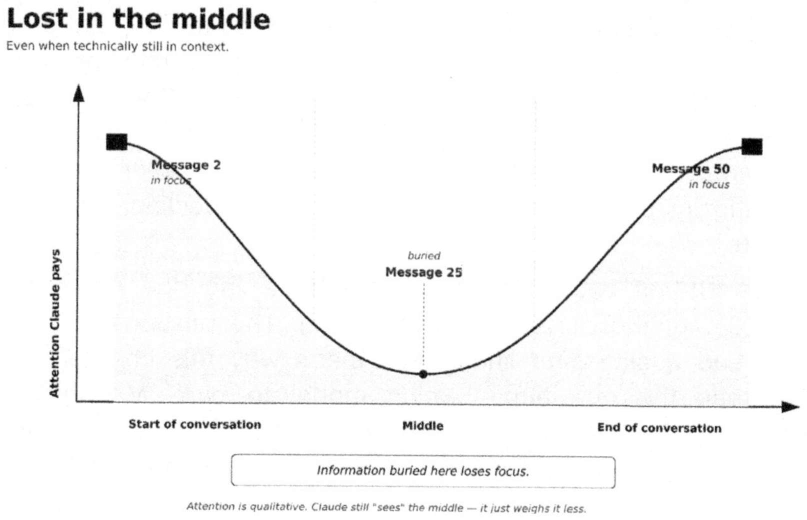

Figure 5.2: The lost-in-the-middle effect. Claude pays more attention to what's at the beginning and end of a conversation. Information buried in the middle loses focus, even when it's technically still on the desk.

Shuffled, incoherent documents produced better results than logically structured ones in Chroma's tests. Structured context amplifies the lost-in-the-middle effect because the model treats coherent text as continuous narrative and processes it with less focused attention on individual facts. Shuffled text forces the model to treat each piece independently.

WHAT THE BENCHMARKS ACTUALLY SHOW

The research explains why more space doesn't automatically mean better results. The benchmarks show how much performance you're losing.

I can usually feel the session degrading before I can prove it. You've already seen what this looks like if you've spent time building with Claude. It starts repeating things it already said, contradicting decisions from earlier in the session. The answers get generic where earlier ones were specific. Confidence gives way to hedging.

The "just get a bigger model" logic sounds right until you look at the benchmarks. Opus 4.6, Anthropic's most capable model, supports a million-token context window. On retrieval benchmarks at 256,000 tokens, it scored about 93% accuracy. At the full million, that dropped to 76%. That's a 17-point decline in retrieval accuracy on a model specifically designed for long-context work. Sonnet 4.5, for comparison, scored 18.5% at the same million-token range, making Opus dramatically better at long context than smaller models. But "dramatically better" still means a significant decline from shorter contexts.

Having 40% of your context window remaining doesn't mean you have 40% of the performance remaining. The relationship between space and quality isn't linear, and that's why this chapter matters more than the one about which model to pick. Managing your conversations is the real lever. The discipline I come back to is simple. One task gets one session. If the work is a big build with several pieces, each piece gets its own session. The alternative is watching a single session try to hold every decision I've made that day, and that's exactly how the wall sneaks up on you. The model you choose sets the ceiling for how long you can push before performance slides. Whether you ever hit that ceiling comes down to how you work.

6
The CLAUDE.md File

You open Claude Code the next morning to pick up where you left off. Yesterday you built the intake form for your Chapter 4 tool, settled on Tailwind for styling, organized the components into a clean folder structure, and got the form submitting to the right endpoint. Today you want to add a new field. Simple enough.

You type your request, and Claude suggests inline styles. You already decided on Tailwind and remind it. Then it creates a file in the wrong directory. You explain the folder structure. Then it recommends an approach that contradicts the layout decisions you spent an hour refining yesterday. You stop and realize what's happening. Claude has no idea what you did yesterday. None of it. As far as Claude is concerned, this is the first time it's ever seen your project.

You start explaining what the tool does, how the files are organized, that you're using Tailwind instead of inline styles, that the form submits to a specific endpoint. Fifteen minutes pass before you type anything that moves the project forward.

And the day after that, the cycle repeats.

If this sounds familiar, you're not alone. In surveys and community threads, this comes up more than anything else: "When you return to a project days later, Claude doesn't remember design decisions, page structure, or tone from earlier sessions." People describe it as training a new contractor every morning who has never seen the project. The briefing is identical. The frustration compounds.

Some users have tried to work around it by keeping one long session running, never closing the chat window, hoping that if they just don't end the conversation, Claude will remember. That instinct makes sense, but it fails in two ways. First, as you learned in Chapter 5, quality degrades as context fills up. The longer you keep a session alive, the worse Claude's responses get. Second, when the session eventually closes, whether you meant to close it or not, everything vanishes anyway. You've been pouring water into a bucket with no bottom.

The GitHub issues tell the same story from a different angle. Someone reported memory files that Claude wrote during a session, saved to the project directory, just gone the next time they opened Claude Code. Another lost their entire session history after a routine update. The sidebar showed "No sessions found." These bugs are real, but they're layered on top of something more fundamental. Sessions aren't meant to persist. The forgetting isn't a bug. It's the architecture.

I fought the forgetting for a while. I kept sessions alive longer than I should have, re-explained my folder structure every morning, and wondered why every day felt like starting over. The fix, when I finally admitted I needed one, wasn't a workaround. It was a file.

WHY EVERY SESSION STARTS FROM ZERO

Every Claude Code session starts fresh. When you start a new session, yesterday's conversation doesn't carry over. Claude sees only what's in the current conversation plus whatever files it reads from your project folder, the folder where you opened Claude Code. That context window you learned about in Chapter 5? Every new session wipes it clean.

This isn't a limitation they forgot to fix. It's a deliberate design choice. Anthropic isolates sessions for privacy, so your conversations don't leak between projects. The isolation also helps performance, because loading old conversations into new sessions would eat context space you need for actual work.

The context window is your desk. It holds a certain amount of material before things start getting buried. A filing cabinet stores things long-term. Claude's session has no filing cabinet built in, just the desk.

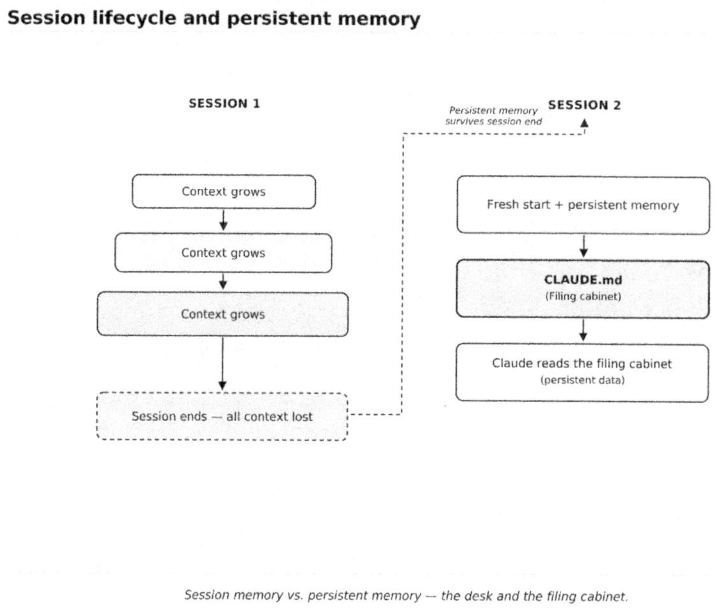

Figure 6.1: Session memory vs. persistent memory: the desk and the filing cabinet.

Geoffrey Huntley, an engineer at Sourcegraph, tested this directly. He pushed Claude Sonnet's context window to find where quality actually drops off. It turned out to be around 147,000 to 152,000 tokens, 25% below the advertised 200,000-token limit. You don't hit a cliff at the end of the context window. Quality erodes well before that. And longer sessions make the erosion worse, because you're adding tokens and Claude's attention spreads thinner across everything it's trying to hold at once.

Your instinct, keeping one long session alive, is wrong twice. The session degrades from context overload while it's running. And everything disappears when it closes. Fresh sessions are the right approach. But fresh sessions create the briefing problem you just read about.

Which means you need a way to carry forward what matters without stuffing it into the conversation.

CLAUDE.MD: YOUR PROJECT'S PERMANENT BRIEF

`CLAUDE.md` is a plain markdown file that lives in your project's root folder, the top-level directory of your project, the same place you open Claude Code from. When you open Claude Code in a project that contains a `CLAUDE.md`, Claude reads it automatically at the start of every session.

That verbal prompt you've been giving every morning? It becomes a file that gives itself. You write down the rules once, and Claude loads them fresh every time.

Imagine you hired a contractor to work on your house. Every morning, they show up and you hand them a one-page brief: what the project is, what's been decided, which rooms are off-limits, what style you're going for. The contractor reads it and gets to work. They don't remember yesterday, but they don't need to, because the prompt covers everything they need to know. `CLAUDE.md` works the same way, except you only write it once, and Claude reads it automatically. For a simple tool, a `CLAUDE.md` might be fifteen or twenty lines. Mine has grown bigger because I use Claude Code across a lot of projects, and the top-level file mostly exists to point Claude at the right rules for whatever I'm working on that day. Either shape works. What matters isn't the length. It's that every line in the file is doing something useful.

For your Chapter 4 tool, a `CLAUDE.md` might look something like a couple of sentences about what the tool does and who it's for, a note that you're using Tailwind for styling, a line about the folder structure, and the commands to run and test the project. Fifteen seconds for Claude to read, instead of fifteen minutes for you to explain.

Claude Code actually has two memory tools that work together. `CLAUDE.md` carries rules you wrote deliberately, things you decided in advance about how the project should work. Auto Memory is different. It lets Claude write notes to itself across sessions: patterns it noticed in your code, commands that worked, debugging approaches that helped. If `CLAUDE.md` is the brief you'd give a contractor on day one, Auto Memory is the field notebook the contractor keeps as they work. You don't need to manage Auto Memory actively. It fills itself in. But knowing it exists helps you understand why Claude sometimes seems to "remember" things you didn't explicitly write down.

There's a cost, though. `CLAUDE.md` loads into the same context window you learned about in Chapter 5. Every line in your `CLAUDE.md` eats tokens that could otherwise be used for conversation, code, and file content. A 500-line `CLAUDE.md` takes up real space, and that space doesn't come back until you start a fresh session. You want Claude to know everything about your project, but everything has a token cost.

THE INSTRUCTION TRAP

You discover `CLAUDE.md` and you do the obvious thing. You sit down and try to capture everything. Every style rule, every file path, every design decision, every pattern you want Claude to follow, every mistake it made last week that you want to prevent this week. Your `CLAUDE.md` hits 500 lines and you feel thorough.

Then Claude starts ignoring half of it. I've done this. The first few `CLAUDE.md` files I wrote were bloated. Every rule I could think of, every preference, every correction from a bad session, all of it in one file. Adherence got worse, not better. The files I write now are shorter, and they work.

You notice it using inline styles even though you wrote "use Tailwind." It creates files outside the folder structure you documented. You catch it hallucinating a test command you specifically said doesn't exist. Your rules are right there, in black and white, and Claude is walking past them.

You do what feels logical. You add more rules, make the existing rules more emphatic, and capitalize IMPORTANT alongside "YOU MUST" in front of the things that matter most. Those emphasis markers do work for one or two critical rules. But when half your file is shouting, none of it stands out. The file grows. And adherence drops further.

Research from HumanLayer tested how many instructions frontier language models can reliably follow at once. The answer is roughly 150 to 200. Claude Code's own system prompt, the instructions Anthropic bakes in before you write anything, uses about 50 of those slots. That leaves you 100 to 150 instructions before reliability starts to slip.

The decline isn't sudden. It's linear. Every instruction you add makes every other instruction slightly less likely to be followed. At 200 rules, Claude follows most of them. At 500, it follows maybe 80%. That sounds reasonable until you realize 80% of 500 means 100 rules being ignored every session, and you have no way to predict which 100.

The pattern is the same one you see when managing people. If you hand a contractor a 30-page brief, they skim it. They read the first page, glance at the rest, and do their best. If you hand them one page with the five things that matter most, they follow it. Claude behaves the same way. The model pays the most attention to instructions at the very beginning and very end of its context. Everything in the middle gets less focus. A massive `CLAUDE.md` means most of your rules land in the attention dead zone.

The trap gets worse over time. After every frustrating session, you add a rule. Claude used the wrong testing framework? Add a rule. It forgot the naming convention? Another rule. Then it created a file in the wrong place, so you add one more. But you never go back and remove rules that no longer apply, or rules that Claude already follows from reading your existing code. The file grows like a to-do list that never gets pruned. Each addition makes every other rule slightly less reliable, which creates more frustrating sessions, which creates more rules.

The `CLAUDE.md` doesn't stop working because of a bug. It stops working because you fed it so much that it can't digest any of it reliably. The fix is fewer, better instructions.

There's also a cost that's less obvious. Claude reads `CLAUDE.md` at the start of every new conversation, which means every line in the file burns tokens every time you open a session. A tight `CLAUDE.md` might cost you a few hundred tokens per session. A bloated one can cost thousands, eating context you could be spending on actual work.

WHAT EARNS A LINE

If 500 lines is too many, what actually belongs in your `CLAUDE.md` ?

Cover what the tool does and who it's for. Without that context, its guesses get further from reality the more complex your project becomes. When Claude knows your tool is a client intake form for a landscaping business, it won't suggest features that make sense for a social media app. That alignment prevents wrong turns before they happen.

Next come code style and patterns, specifically the constraints Claude can't reliably infer from a small codebase. "Use Tailwind utility classes, not inline styles." "File names use kebab-case." "Each component gets its own folder with a matching test file." For your Chapter 4 tool, this might be three lines. These earn their place because Claude adapts to patterns it sees in existing code, but when your codebase is small or inconsistent, explicit direction prevents drift.

Then, add your exact build, test, and run commands. " `npm run dev` " starts the development server. " `npm test` " runs tests. Include the flags. This prevents Claude from hallucinating commands that don't exist in your project. Beginners lose real time debugging commands Claude confidently invented, commands that look plausible but fail because your project's setup doesn't include them.

Finally, include a short map of key files and what they do. Claude sees your entire codebase, but it doesn't know which files matter most. A three-line map, "src `/components` / holds React components, src / `lib` / has utility functions, public/ has static assets," gives Claude a starting point without requiring it to read every file before making a change.

That's it for a project the size of your Chapter 4 tool, maybe 20 lines total. You can write it in five minutes, and it'll save you fifteen minutes every session.

You don't even have to write it by hand. The fastest way to get a `CLAUDE.md` is to ask Claude Code to make one for you. Open your project folder, then type something like: "Read the files in this folder and create a `CLAUDE.md` that captures what this tool does, the styling I'm using, the folder structure, and the commands I run to test and ship it." Claude reads what's already there, drafts the file, and shows it to you before saving. You read it the way you'd read a brief from a contractor. Keep what's right, cut what isn't, add the one or two things only you know. The file lands in your project root and starts loading on the next session.

A working target is roughly 200 to 400 words for most projects. If your `CLAUDE.md` is creeping past 600, that's a signal. The file is doing too much, and adherence is about to drop. Appendix B walks through how to diagnose this when Claude starts ignoring rules you wrote down.

Here's what a lean `CLAUDE.md` for the Chapter 4 intake form looks like:

```
# Client Intake Form

A simple HTML/CSS/JS form for collecting freelance project
inquiries. The
submission saves to data.json so I can review entries
between client calls.

## Style and patterns
- Plain HTML/CSS/JS, no framework
- Inline form validation (no submit until required fields
are filled)
- Class names use kebab-case
- Keep the look professional and minimal, since this lives
on a freelance site

## Files
- index.html: the form markup
- styles.css: layout and styling
- script.js: validation, submit handler, file write
- data.json: where submissions land

## How I use it
- Open index.html in a browser to test
- Each new submission appends to data.json as one JSON
object
- I read the file by hand for now; export comes later

## Things to avoid
- No backend server yet
- No CDN dependencies, this should keep working offline
- Don't add fields without asking; the form is intentionally
short
```

Eighteen lines. Five minutes to write. Claude reads it on every fresh session and saves you the briefing.

The most important principle is how the file grows from here. Don't try to write the perfect `CLAUDE.md` before you start building. Start with those 20 lines. Then pay attention to what goes wrong. The same goes for folder structure.

The best `CLAUDE.md` files aren't written in one sitting. They're grown from corrections. Each rule earned its place because Claude did the wrong thing without it. Every line in the file answers the question "would removing this cause Claude to make a mistake?" If the answer is no, the line doesn't belong there.

Rule of thumb: if removing it wouldn't cause Claude to make a mistake, cut it.

Include in CLAUDE.md	Don't include
Project structure (key folders and what they hold)	Full file contents — link to them instead
Testing framework you use (e.g. "always use Vitest, not Jest")	General coding best practices — Claude already knows these
Naming conventions for your project	Things you figured out once and never need to repeat
Packages to prefer or avoid (and why)	Database schemas — point to the schema file
Rules that fix recurring mistakes Claude makes	Anything that could be in source code comments
A three-line map of where things live	Your whole README

Table 6.1: What earns a line in your CLAUDE.md (and what doesn't)

As your project grows, you can keep the file lean by pointing Claude to information instead of copying it in. Instead of pasting an entire database schema into `CLAUDE.md`, write one line: "For database schema design, see docs/`DATABASE.md`." That takes up one line instead of fifty, and Claude can read the referenced file when it needs the information.

That pattern shows up everywhere builders use Claude Code. The people who get the strongest results aren't writing the most technical `CLAUDE.md` files. They're the ones who understand their domain deeply enough to give Claude the context it can't find anywhere else.

Consider a bookkeeper who describes exactly how expense categorization works in their industry, a therapist who specifies the clinical language constraints for patient-facing tools, or a contractor who knows which permit fields California municipalities actually check.

`CLAUDE.md` is advisory, not deterministic. Even a well-written file lands somewhere around 80% adherence. That's the right tool for guidance at this stage of your project, but it's worth knowing before you rely on the file too heavily.

THE BUILDER WHO STOPPED FIGHTING IT

J.D. Hodges is an attorney and IT professional. He spent eight sessions over multiple days getting a self-hosted Open WebUI and SearXNG stack running on a Proxmox server. Configuration values discovered in session three, IP addresses, file paths, service-specific settings, were still relevant in session eight. Without a system for preserving that information, every new session started the same way: minutes of re-orientation, re-explaining the architecture, re-establishing what had already been tried and what had already failed.

His solution was two files. The first, a project reference file, held permanent information: the architecture, IP addresses, file paths, and service configurations that wouldn't change between sessions. The second was a living session log with dated entries recording what was done, what was tested, what worked, what didn't, and what needed to happen next.

At the end of each session, Hodges had Claude generate a structured handoff: the current goal, the status of each component, decisions made, things to avoid, open questions, and the specific next step. The new session loaded this handoff and picked up where the last one left off. No catching up, no watching Claude suggest an approach that failed two sessions ago.

The shift wasn't technical. Hodges changed how he thought about the tool. He stopped treating Claude like a continuous collaborator who remembers yesterday. He started treating it like a brilliant colleague with amnesia. Every session starts from zero. The question isn't whether Claude will remember. It won't. The question is what you put on the desk before it begins.

Decker, a developer who wrote about his approach publicly, tried three different memory systems before landing on one that worked. He tried structured notes, a database layer, and session replay. Different solution, but the insight is the same.

Hodges used `CLAUDE.md` to its full weight. Alongside it lives a session log, referenced from `CLAUDE.md` with one line. "See `SESSION_LOG.md` for what's been tried and what's open." When Claude starts a new session, it reads `CLAUDE.md`, sees the pointer, pulls in the log, and picks up where the last session ended. The work scales because the file system does.

I have a version of this rule in my own setup. When I run an autonomous build, the kind where I hand Claude a prompt and walk away, the first step in the prompt tells Claude which rules to read. I learned to write it that way after one of those builds skipped the step and produced work I had to throw out. The forgetting isn't a flaw. It's what the system does. Design for it and the system works for you.

You don't need to build Hodges' setup on day one. A weekend project won't need it either. The first time a project spans more than a handful of sessions and you catch yourself re-explaining the same decisions, the pattern is waiting. Open `CLAUDE.md`, add a pointer to a second file, and keep working.

7
Plan Mode

You want a client dashboard, something bigger than another tweak to the intake form. A real feature with its own screen showing every submitted intake form, letting you update the status of each one, and connecting back to the original submissions so nothing falls through the cracks. It needs a backend component to pull the data, new interface elements to display it, and some way to track which submissions are new, which are in progress, and which are done. This is the kind of thing that would've cost you $5,000 to hire out six months ago.

Adding a small feature to a working project is usually simple. You describe it, Claude builds it, you adjust a detail, done. Adding something like this dashboard is where builds go sideways. Not because the feature is too hard for Claude. Because a complex addition touches so many files at once that every decision Claude makes in one place creates a conflict somewhere else. Without a plan for how the pieces connect, the time you spend building is time spent cleaning up conflicts.

You do what's worked every time so far. You open Claude Code and describe what you want. Claude starts building immediately. Files appear, new components and new routes and what looks like a data layer you didn't ask for but that Claude apparently decided you needed. Twenty minutes in, you notice something wrong. Claude chose a different data structure for the dashboard than what your intake form already uses. The form stores submissions one way. The dashboard expects them another way. They don't talk to each other.

You explain the problem. Claude apologizes and starts refactoring. It rewrites the data layer to match your form's structure. But now the original form submission is broken. Something in the refactoring changed a file your form depends on, and submissions that were working ten minutes ago throw an error you've never seen before.

An hour in, you have more files in your project than when you started, and you're not sure what half of them do. You're spending all your time fixing things that worked before you touched them. The dashboard feature is maybe 30% done, and every correction you make seems to introduce a new problem somewhere you weren't looking. You started the morning wanting to add a feature. You're ending it wondering if you broke your entire project.

This isn't a Claude Code problem. It's a direction problem. Nobody planned which pieces connect where before the building started. Nobody made the architectural decisions on purpose. Nobody decided on purpose how data would flow between components, where new files would live, or which patterns to follow. All of that got decided implicitly by whoever typed first. In this case, Claude typed first. And Claude made reasonable guesses that happened to conflict with what already existed.

Every correction you feed Claude goes into the same conversation context, and that context gets noisier with every fix. Anthropic's own best practices identify this exact pattern, what they call "correcting over and over," as one of the most common failure modes. After two failed corrections, they recommend clearing the conversation and writing a better initial prompt instead of continuing to patch. A cluttered context makes each subsequent correction less reliable than the last.

The vibe coding community calls this the 80/20 wall. AI-generated code works beautifully for the first 80% of a project. The initial build works. The basic structure holds. The simple features that don't touch each other come together fast. Then you hit the final 20%, the edge cases, the integrations between components, the production hardening, and suddenly the project requires exactly the skills these tools promised you wouldn't need. A late 2025 analysis documented this pattern across multiple communities, and it resonated because everyone building with AI had lived some version of it.

The most-engaged thread in r `/vibecoding` tells the same story from a different angle. "The problem with vibe coding is nobody wants to talk about maintenance," the post title read. It got 562 upvotes and

252 comments. The problem resonated because it wasn't about code quality. It was about what happens when you build without a plan and then try to change what you built. Every modification cascades into something adjacent. The code itself is fine. The project becomes fragile because nobody mapped out how the pieces relate to each other before the building started.

Someone in those same communities spent $100 over two days arguing with Claude Code on a single task. Same conversation, same problem, going in circles. Then they solved it in ten minutes by backing up, breaking the task into two smaller pieces, and starting fresh. The approach was the whole problem.

This is what happened with your dashboard. You didn't need better code. You needed a plan, a way to understand what your project already had, decide how the new feature would connect to it, and agree on an approach before Claude started creating files. The thinking before the building was the part you skipped.

READ BEFORE YOU BUILD

Two contractors. Same job. The first one shows up at your house, listens to you say "I want to remodel the kitchen," and immediately starts tearing out cabinets. He's fast. He's confident. And by the end of the day, you discover he ripped out the wall where your plumbing runs because nobody checked the blueprints first.

The second contractor walks through your kitchen with you before touching anything. She opens cabinets, checks where the pipes run, looks at the electrical panel, asks you questions about what you cook and how you use the space. She hasn't built anything yet, but by the time she picks up a tool, you both know exactly what's happening and why.

Claude Code has a mode that turns it into the second contractor. It's called Plan Mode, and the idea is deceptively simple. It lets Claude read everything in your project but change nothing. Claude can open files, search your codebase, analyze how your components are structured, examine your routing patterns, read your `CLAUDE.md` , and pull information from documentation. What it can't do is edit a file, create a file, delete a file, or run any command that modifies anything. That read-only constraint is the whole point.

Anthropic's own best practices documentation puts it plainly: "Letting Claude jump straight to coding can produce code that solves the wrong problem. Use Plan Mode to separate exploration from execution." The recommended workflow has four explicit phases: Explore, Plan, Implement, Commit. Plan Mode handles the first two.

This is the Chapter 3 skill scaled up. Context engineering told you how to close decisions before Claude starts building. For small tools like the intake form, you closed them inside a single prompt and a few rounds of clarifying questions. For features like this dashboard, the decisions are too many and too interdependent to fit in a brief. Plan Mode gives you a dedicated space to work them out, with Claude able to read everything in your project and unable to touch any of it.

You activate Plan Mode by pressing `Shift+Tab` twice. This cycles through Claude Code's permission modes until you land on Plan. You can also type the `/plan` command, or start a session with the --permission-mode plan flag if you know you want to plan before you build. Once Plan Mode is active, you'll see a visual indicator that Claude is in read-only mode. There's no ambiguity about what state you're in.

`Shift+Tab` cycles through three modes by default: default, acceptEdits, and plan. Auto mode is also available but requires opt-in before it shows up in the cycle. You'll spend most of your time in default and plan. The others give Claude more or less autonomy over which actions it can take without asking you first, but for now, the two that matter are default mode for building and Plan Mode for thinking.

In default mode, you can ask Claude to "look at my code first before making changes," and sometimes it will. But it can also start editing at any point during its analysis. You've probably experienced this already. You ask Claude a question about your project, and instead of just answering, it starts modifying files because it saw something it thought it could improve. Plan Mode prevents that structurally. You don't need willpower to keep Claude from jumping ahead. The mode handles it.

When you're ready to start building, press `Shift+Tab` again to cycle back to default mode. Claude regains the ability to edit files and run commands, and you'll see the visual indicator change to confirm you're out of Plan Mode.

Armin Ronacher, the creator of Flask (one of the most popular web frameworks in Python), dug into how Plan Mode works under the hood. He found it operates through a system reminder that tells Claude: "Plan mode is active. The user indicated that they don't want you to execute yet, you MUST NOT make any edits." That simplicity is the point. The constraint works because it's absolute, not because it's technically complex.

Plan Mode is where you ask the questions you've never thought to ask about your own project. "How is data currently stored?" "What pattern do the existing components follow?" "If I wanted to add a dashboard, what would need to change?" Claude reads your codebase and answers from what it sees, not from generic knowledge about how projects usually work. The answers are specific to your project, your file structure, your existing patterns.

If you want to edit the plan Claude produces, press `Ctrl+G`. This opens the plan in your default text editor so you can make changes directly before Claude proceeds. You're not limited to describing changes in the chat. You can write on the plan itself.

Plan Mode isn't only for big features. Anthropic recommends it whenever the approach is uncertain, when changes would touch multiple files, or when you're working with code you haven't seen before. A five-minute planning session before a moderate change can save you from the same Circle of Doom that eats entire afternoons on bigger features.

PLANNING A FEATURE TOGETHER

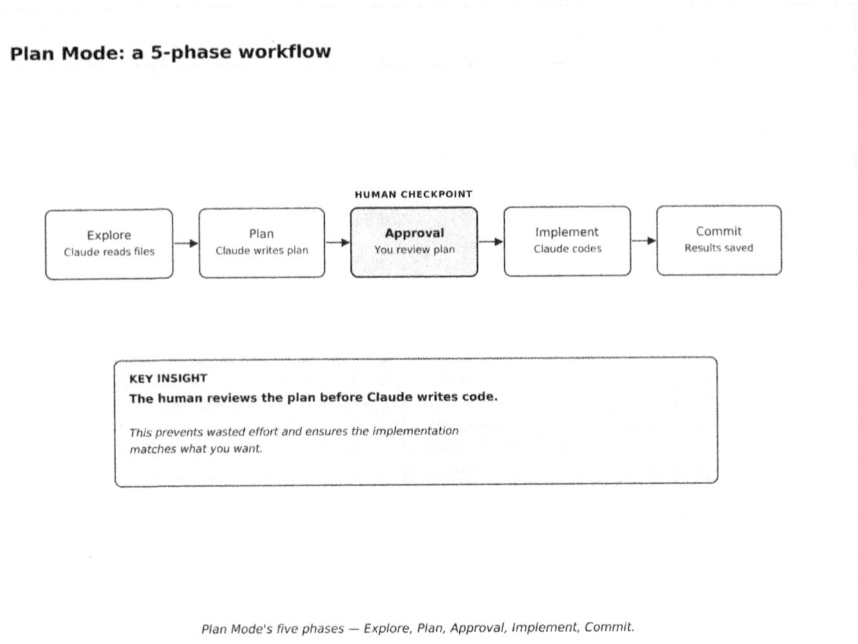

Figure 7.1: Plan Mode's five phases: Explore, Plan, Approval, Implement, Commit.

Let's take the same dashboard feature that went sideways earlier and build it the right way. Same project, same goal, completely different process. We're going to walk through all five phases using your actual Chapter 4 tool as the starting point. Anthropic's framework has four. I add a fifth, Approval, between Plan and Implement, because the plan shouldn't ship without your sign-off. The full sequence: Explore, Plan, Approval, Implement, Commit.

Start by activating Plan Mode. Press `Shift+Tab` twice, or type `/plan`. Then ask Claude to read your project files.

```
Read through the project and tell me how it's currently
structured. Where does the intake form store data? What
components exist? What routing pattern are we using?
```

Claude goes through your codebase and reports back. It tells you the intake form stores submissions in a specific file or database format, maps out which components handle which parts of the interface, and shows you the routing pattern your project follows.

Ask follow-up questions. "Where do API calls happen?" "What would need to change to display submitted forms on a new page?" "Is there anything in the current structure that would make adding a dashboard difficult?" Each question costs a few tokens, but the answers prevent wrong directions that would cost you hours of rework. Heeki Park, an AWS Principal Solutions Architect, described spending "a lot of time in the planning phase" before writing any code with Claude Code. His follow-up interactions during building were "small tweaks rather than wholesale changes to the entire project." The time he spent exploring and planning wasn't wasted time. It was the reason everything after it went smoothly.

Once you understand what you're working with, describe the dashboard feature in plain language. Technical jargon won't help here.

```
I want a new page that shows all the intake form submissions. Each one should show the client name, the date it was submitted, and a status field I can update. When I change the status, it should save without refreshing the page.
```

That's enough for Claude to work with.

Ask Claude to propose an implementation plan. Claude breaks the feature down into which existing files need to change, what new files are needed, what order to build them in, and what edge cases it can already see. You push back. "I don't want to change the database schema, can we work with what's already there?" "Can we keep the existing form route untouched so nothing breaks?" "What happens if someone submits a form while I'm looking at the dashboard?"

Three to five rounds of this back-and-forth turn a generic implementation plan into one that fits your project, your constraints, the priorities you care about. You build the plan together. You're making architectural decisions here, deciding where data lives, how components talk to each other, what the user experience should feel like, even though you're not writing a single line of code.

This back-and-forth is the Approval phase. You're reviewing the plan, pushing back on the parts that don't fit, and shaping it before Claude writes any code. The plan is approved when you can read it through and trust that it does what you want with the constraints you have. The dashboard mess earlier in the chapter happened because nobody pushed back on the plan. The time you spend here is what keeps the next one from showing up.

Anthropic recommends that once your plan is solid, you start a fresh session for implementation. The planning conversation served its purpose, and starting clean gives Claude a focused context with just the plan and your `CLAUDE.md` , instead of twenty rounds of back-and-forth exploration cluttering its memory.

Before you switch sessions, ask Claude to save the plan you just built. A simple prompt does it.

Save what we discussed to `FEATURE_PLAN.md` in this folder.

Claude writes the file. Read it the way you'd read a brief from a contractor. Keep what's right, cut what isn't, and add the one or two things you know that Claude doesn't.

Then start the fresh session. The simplest move is to type `/clear` , which empties the current chat without leaving your project folder. (You'll see `/clear` again in Chapter 8, the move you make whenever a session is fighting against you.) Claude reads your `CLAUDE.md` automatically when the chat clears. Tell it to read `FEATURE_PLAN.md` and start with the first chunk.

Now you switch out of Plan Mode. Claude can make changes again, but this time it's working from a shared plan instead of guessing at architecture. You build in small slices. First you tackle the data display, checking whether the dashboard can pull and show the submitted intake forms. That works, so you move on to the status update, testing whether you can change a submission's status and have it save. Once that's solid, you wire up the connection between the form and the dashboard, confirming that new submissions appear without a manual page refresh.

Each slice is small enough that if something breaks, you know exactly which change caused it. Compare this to the approach that went sideways earlier, where Claude built everything at once and you couldn't tell which file was responsible for which problem. Small slices are faster because you don't lose time debugging cascading failures.

After each working slice, save a checkpoint. If you've set up version control for your project, this means committing your changes. If not, the concept is the same. Preserve the current working state so you can get back to it if the next slice breaks something. Rolling back to the last good state is always faster than trying to untangle what went wrong in the middle of a broken build.

When the feature is complete, that plan you saved earlier becomes part of your project's memory, sitting alongside `CLAUDE.md`. If you close Claude Code and come back to this project in a week, the plan is still there. Claude reads it and picks up the reasoning behind what you built.

Park documented what changed once he started planning before building. Claude's output quality stayed the same. Everything after the first build got better. Follow-ups became small tweaks instead of wholesale rewrites. The architecture held as features were added. Code that worked on day one still worked on day thirty. Planning didn't slow the build down. It made the build stick.

Builders in the vibe coding community report similar results. Those who follow spec-driven approaches, where the plan is written down and reviewed before any code gets generated, report 80 to 90% fewer bugs during implementation. That's community-reported, not peer-reviewed, but the pattern is consistent enough across hundreds of posts that the direction is clear.

WHAT GOES IN THE PLAN

Before you build anything, you need three things written down. What the feature does, in plain language that you can verify without reading code. What the feature doesn't do, the explicit list of things that are out of scope. And how the feature connects to what already exists, which files change and which stay untouched.

The first part is what most people get right naturally. "The dashboard displays all submitted intake forms with their status. I can click on a submission to see the details. I can update the status from 'new' to 'in progress' to 'completed.'" That's plain language, clear expectations, and a test that doesn't require reading code. You can see it working or you can't.

The second part, the out-of-scope list, is what most builders skip, and it's the part that saves the most time. Write it as a list of explicit negatives: "This dashboard does NOT add user authentication, does NOT change the intake form itself, does NOT add email notifications, and does NOT create a database migration." Every item on that list is something Claude might have built if nobody told it not to. You saw this earlier in the chapter. Claude decided your project needed a new data structure nobody asked for, and that decision broke the existing form.

The anti-drift workflow guide nails it. The out-of-scope section is "your reference point when you're tempted to add 'just one more feature.'" It's not saying no to good ideas. It's saying "not yet" so the current feature ships. Every feature you don't explicitly exclude is a feature Claude might include, and every unplanned addition is a potential conflict with something that already works.

The last piece is about how the new feature fits into what's already there. Which files will Claude need to modify? Which should it leave alone? Where does the new code live in the project structure? If your intake form stores data in a JSON file, the plan should say the dashboard reads from that same file, not from a new one Claude might invent. If your project uses a specific component pattern, the plan should say the dashboard follows that same pattern.

You might be thinking you need to understand the code before you can write a plan like this. You don't. That's what the Explore phase is for. Claude reads the codebase and tells you what's there, which means the plan you write together is grounded in the actual project structure, not in assumptions you'd have to make on your own. Your job during planning isn't to understand every line of code. It's to ask the hard questions about what you want the feature to do and how it should behave, because Claude might not ask those questions unprompted. The domain expertise, knowing what your clients need and how your business works, is yours. The technical translation is Claude's.

Once you have those three pieces, break the feature into small testable chunks. The dashboard isn't one task. It's three or four: display the data, add status updates, connect the form submission to the dashboard view, style the layout. Each chunk has a clear "done" state you can verify without reading a line of code. If the submitted forms show up on the dashboard with the right names and dates, chunk one is done. You don't need to understand the code to know that.

Between chunks, I run what I call a completion analysis. Before starting the next piece, ask Claude to review what was built against the plan.

```
Does the dashboard implementation match what we planned? Did anything change that we didn't plan for?
```

This catches drift before it compounds. One deviation in chunk one becomes three deviations by chunk three if nobody checks. The review takes thirty seconds. The rework from unchecked drift takes hours.

Anthropic's best practices describe a more structured version of what Chapter 3 already showed you. Give Claude a brief feature description and ask it to interview you, working through implementation details, edge cases, and tradeoffs by asking clarifying questions until the spec is tight enough to build from. You answer. The result is a spec written to a file that both you and Claude can reference throughout the build. Same move you learned for small tools, formalized to handle features that need to survive multiple sessions.

WHEN CLAUDE ASKS PERMISSION

The first time Claude wants to change a file or run a command, it pauses and asks. You'll see a prompt in your chat window that names the action. You can approve it once, approve it permanently for this command, or refuse and tell Claude what to do differently.

Read the action before you decide. Edit prompts show you the file Claude wants to change. Bash prompts show you the exact command. When Claude wants to use a connected tool (read your Google Calendar or send a Slack message, for example), the prompt shows which connector and what it's about to do.

The decision rule is simple. Did you ask Claude to do this? If you said "build a contact form" and Claude asks to install Express, that matches what you asked for, so approve. If you said "rename this file" and Claude asks to delete the entire folder, that doesn't match. Say no and ask Claude what's going on.

When something looks unfamiliar, type a question.

```
What does this command do? Why are you running it?
```

Claude explains, and then you decide. The prompt waits for you. Taking ten seconds to think doesn't slow anything down.

Once you've approved a command a few times and you trust it, pick the second option (allow always for this command). The prompts stop for that specific action and keep coming for everything else. You don't have to do this for the whole tool, only for the commands you've already vetted.

If you want read-only access to your project files, switch back to Plan Mode. You've already seen how it works.

When you want finer control, type `/permissions` . It lets you set rules per command pattern so commands you've vetted run without asking and ones you've banned never run. You won't need this on day one. When you do, Claude Code's documentation walks through the syntax.

These prompts exist to protect you. Saying no to something that surprises you costs nothing. Approving everything to make the prompts go away gives away the protection.

Every chapter that follows builds on this loop. The projects get more complex and the prompts get more specific. But the interaction pattern stays exactly the same.

8

The Hidden Costs Within Claude Code

You can burn through a full day's quota in an hour without noticing. The meter inside Claude Code doesn't flash red. It quietly logs every message, every file read, every tool call against a budget you probably never looked at. This chapter is about that budget. Session habits end up costing more than the plan you picked, and most builders don't figure that out until they've paid for it.

Pro is one of four tiers. The Free plan gives you chat access to Claude but not Claude Code itself, so you can't build on it. Pro at $20 a month is where building starts. Max 5x at $100 and Max 20x at $200 exist for people who code heavily every day and burn through the Pro quota regularly. I use the $200 Max 20x plan because I use Claude Code daily to build and run systems for my businesses. For anyone working through this book, Pro is the right choice. Max gives you more usage before a limit kicks in, not better AI.

Plan	Monthly cost	Claude Code access	Best for
Free	$0	Chat only, no Claude Code	Trying Claude; not for building
Pro	$20	Yes — flagship models, extended thinking, context up to 1M tokens	Everyone reading this book. The starting plan.
Max 5x	$100	Yes — 5× the Pro quota	Heavy daily users who hit the Pro ceiling
Max 20x	$200	Yes — 20× the Pro quota	Full-time builders using Claude Code every day

Table 8.1: Claude Code pricing tiers. Pro is the right starting plan for anyone building their first project.

WHAT YOU'RE ACTUALLY PAYING FOR

There's also an API option where you pay per token instead of a flat monthly fee. "Pay only for what you use" sounds cheaper. It's almost never cheaper for the kind of interactive building this book teaches. One developer tracked eight months of daily Claude Code usage, roughly ten billion tokens of work. At API rates, that would have cost over $15,000. On the Max plan, the actual bill was around $800 total. The peak month alone would have cost $5,623 at API rates, more than four and a half years of a Max 5x subscription.

The gap is that dramatic for a specific reason. Over 90% of the tokens in a typical Claude Code session are Claude re-reading your project files to stay oriented, also known as cache reads. Your subscription covers those cache reads. On the API, you pay for every one of them.

The real comparison isn't subscription versus API, though. It's $20 a month versus the $15,000 to $60,000 it would cost you to pay a developer to build an app or business automation for you. For the cost of a single lunch out, you get the tool that replaces the quote that made you close the email.

"Pay only for what you use" feels frugal. For the interactive, back-and-forth building this book teaches, subscription pricing is dramatically cheaper. The developer who tracked eight months of usage would have paid 19 times more on API than on Max. For a beginner on Pro, the math is even more lopsided. You've spent more than $20 a month on SaaS tools that solved less of your problem.

HOW USAGE BECOMES COST

Your subscription covers the heavy lifting, but it has edges. Quotas work on two timers. There's the five-hour rolling window that limits how many prompts you can send in a burst, and there's a weekly rolling cap that limits total usage over seven days. On Pro you get roughly 45 prompts per five-hour window, which is plenty for focused work. Anthropic's own data shows the average Claude Code user costs about $6 per developer per day, with 90% of users under $12 a day. For a beginner building their first project, you'll be well under that ceiling.

If you find yourself hitting the quota wall more than once a week, that's the signal to consider Max 5x at $100 a month. You'll know because Claude will tell you that you've reached your limit and need to wait for the window to reset. Until that happens regularly, Pro is the right choice.

When you do hit a limit, it isn't subtle. Claude pauses and tells you directly that you've reached your usage for the current window. There are no surprise charges and no mystery deductions. You either wait for the window to roll over or enable extra usage to keep going at per-token rates.

That overflow is where a subscription plan can start to feel like a bill. Extra usage kicks in at per-token API rates, so you can keep working when you hit the wall, but you pay more for those messages than you would under your normal quota. You're never truly stuck on a paid plan, but you're also not locked at a flat rate once the quota is spent. You still stay in control of the spend, though. You pay for overflow in advance, and you can set a monthly cap on API usage.

One more wrinkle. Your Claude Code usage, your Claude.ai chat usage, and your Claude Desktop usage all draw from the same pool. If you spend your morning having long conversations in Claude.ai chat, you'll have less Claude Code quota available in the afternoon.

The subscription gives you a generous ceiling. But how you use Claude Code inside that ceiling still matters, because bloated sessions burn through quota faster, and quota you've burned through gets expensive quickly.

A developer sat down one morning with a bug that should have been simple. A function wasn't returning the right value. The data looked correct going in, but something broke on the way out.

They opened Claude Code and started describing the problem. Claude suggested a fix. It didn't work. They explained why it didn't work, pasted in more code, described the behavior they were seeing. Claude suggested something else. That didn't work either. They went back and forth, adding context each time, sharing error messages, trying Claude's suggestions, reporting back when each one failed.

Two days later, they'd spent $100 on that single conversation. The bug was still there.

Then they tried something different. They closed the conversation, opened two fresh ones, and split the problem in half. In the first conversation, they asked Claude to analyze just the data transformation. In the second, they focused on just the output formatting. Ten minutes later, the bug was fixed.

The bug didn't cost them two days and $100. The conversation did.

Claude processes every message you send against everything that came before it in the session. The first message is cheap. The twentieth message carries the weight of the previous nineteen. By message fifty, Claude is re-reading an enormous amount of text just to process your latest question. The conversation doesn't just get longer. It gets more expensive per message as it grows.

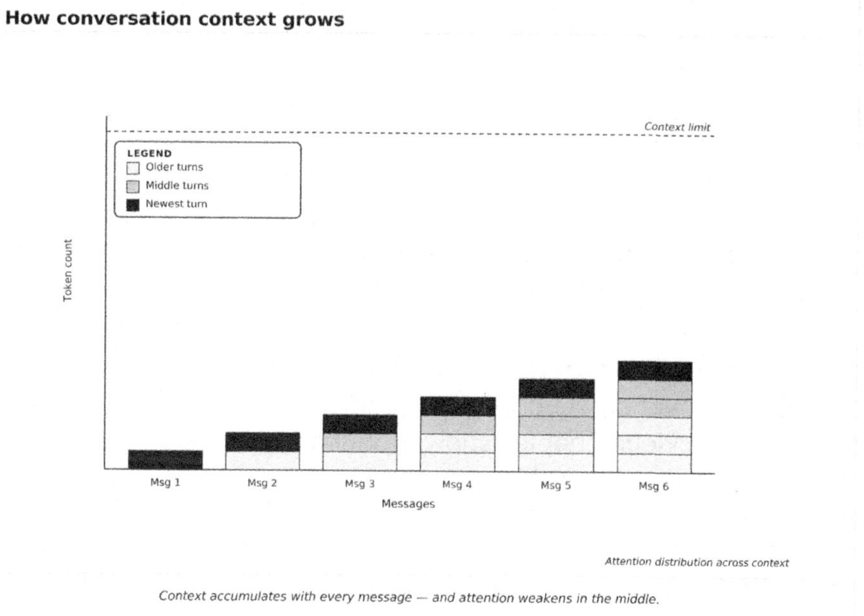

Figure 8.1: Context accumulates with every message, and attention weakens in the middle.

That $100 bill was a workflow failure, not a pricing failure. One very long conversation kept accumulating context without ever resolving the issue. How you use Claude Code determines what it actually costs you.

WHEN TO START A NEW CONVERSATION

Experienced Claude Code practitioners have independently landed on the same rule. Don't let context exceed 60 to 70% capacity. To check your usage, type `/context` in any session. Claude Code shows you how much of your window you've used, broken down by what's taking up the space. When that number climbs past the midpoint, it's time to think about starting fresh, because you're protecting the quality of what comes next and the quota that pays for it.

If you'd rather see usage at a glance instead of typing `/context` every time, you can configure a permanent indicator. Run `/statusline` once and tell Claude what you want to display. Something like `/statusline` show context percentage and model gives you a status line at the bottom of your window from then on. With it set up, you can watch your desk fill up in real time without typing anything.

Five situations make starting a new conversation faster than continuing. You've just finished a feature, fixed a bug, or completed a major milestone. Fresh context for the next job beats carrying the history of the last one. You've corrected Claude more than twice on the same issue. The context is now full of failed attempts, and each correction adds tokens that push you closer to the wall. A single, refined prompt in a clean session will outperform continuing through accumulated confusion.

Your context has crossed 60% capacity. You might have space left, but as the previous chapter showed, remaining space and remaining quality aren't the same thing. You're switching to an unrelated task, like building a frontend and then pivoting to database setup. The old context actually hurts because Claude has to reason across two unrelated domains. Or you're returning after hours or days away. Long idle gaps mean you've lost the session's reasoning thread anyway. A fresh start with good notes is cleaner than a stale conversation.

Three situations make staying in the same conversation the better call. You're iterating on a single feature, making small changes, testing, refining. Continuity helps here because Claude needs to see the evolving state.

Context is still below 60% and Claude is tracking correctly? Keep going. Switching conversations costs time, and if things are working, let them work. Same goes if you haven't corrected Claude yet, or only once. The session is healthy. Don't abandon a good conversation to solve a problem you don't have.

Start a new conversation when...	Keep going when...
You just finished a feature, fixed a bug, or hit a milestone	You're iterating on a single feature (small changes, testing, refining)
You've corrected Claude more than twice on the same issue	Context is still below 60% and Claude is tracking correctly
Context has crossed 60–70% capacity	You haven't corrected Claude yet, or only once
You're switching to an unrelated task (e.g. frontend → database)	The session is healthy and things are working
The conversation has gone sideways and you can feel it	Continuity matters for what you're doing next

Table 8.2: Quick reference for when to start a new conversation versus when to keep going.

Before you clear, type `/rename` and give the conversation a name you'll recognize, something like "homepage-layout-v1." That way you can find it later if you need to go back.

THE HANDOFF BETWEEN SESSIONS

Before starting a new conversation, have Claude write a summary of progress and key decisions to a file. Commit your code so you have a safe recovery point. Start a fresh session. Load the summary in the new conversation. You get a full context budget with all the decision history from the previous session, fresh attention, and no lost knowledge.

Compaction is the Claude Code feature behind that handoff move. It compresses your conversation history into a shorter summary that preserves key decisions and context. You trigger it manually by typing `/compact`, or Claude triggers it automatically when you're running out of room. The forced version is the one that bites. When Claude has to compact mid-session because the window is full, it sometimes loses recent work that hadn't been written down yet.

(Appendix B walks through the manual `/compact` pattern and when to reach for it.)

So there's a workaround for the auto-compact problem, refined over several months by a developer named Ben Fisher. A custom command triggers proactively at 70 to 80% capacity, before compaction forces a reset. The command writes a summary of the current goal and saves working details to a context file. On resume, the new session loads both. Fisher's insight is that waiting until things break is the wrong strategy. By the time you notice degradation, you've already lost quality in whatever Claude produced during the decline.

Black Dog Labs, an engineering team, quantified what this handoff saves. Their protocol reduced session handoff context from over 10,000 tokens of manual re-explanation to under 2,000 tokens, an 80 to 90% reduction. The new session starts with a clear, focused prompt instead of a bloated summary that eats into the fresh context budget.

Monday morning, you open a fresh session and load Friday's summary. You build one feature, summarize, commit. Tuesday, you open a new session for the next feature. Each session stays sharp because the context stays focused on one job. The project moves forward across sessions because the handoff captures what matters and discards what doesn't.

The pattern is simple. Before you close a session, have Claude write down what it knows. Before you start the next one, give it back.

THE THREE HABITS THAT KEEP COSTS DOWN

Anthropic's own documentation lists twelve techniques for managing Claude Code costs. Most of them require project experience you don't have yet. Three habits handle the root cause of almost every cost problem beginners encounter, and you already have two of them from the sections above.

The first is starting fresh when you switch tasks or when the session has gotten too long. That discipline is the biggest lever you have.

The second habit is using Plan Mode before you build anything complex. Press `Shift+Tab` twice, and Claude switches from writing code to proposing an approach. It looks at your project, thinks through what needs to happen, and lays out a plan for you to review before Claude writes a single line of code.

This saves money because the most expensive mistake in Claude Code is going in the wrong direction for too long. Without Plan Mode, you might spend twenty messages building an authentication system one way, realize it doesn't fit your project's needs, and then spend another twenty messages rebuilding it differently. Plan Mode catches that mismatch before the first line of code, when fixing it costs one message instead of forty.

The third habit is being specific in your requests. "Improve this codebase" sounds reasonable, but it tells Claude to scan every file in your project looking for things to improve. That broad sweep burns through tokens fast. "Add input validation to the login function in `auth.ts`" tells Claude exactly where to go and what to do.

All three habits are doing the same thing under the hood. Every one of them reduces the amount of context Claude processes per message. Fresh conversations keep that number small. Plan Mode keeps you from going the wrong direction, which is the expensive kind of context because you're paying to undo work. Specific requests just skip the files that don't matter. Stay focused, stay cheap.

9

Connecting Claude Code to Your Real Tools

If someone told you "your tool needs to connect to external services via their APIs," you'd feel like you just hit a wall. API sounds like something developers deal with. It sounds like code, protocols, and technical specifications that have nothing to do with building a client intake form or a quote calculator. That's the reputation, but the concept is one you use every day.

You sit down at a restaurant. The waiter comes over, and you look at the menu. You pick what you want and tell the waiter your order. The waiter takes that order to the kitchen, the kitchen prepares it, and the waiter brings back your food.

An API works the same way. API stands for Application Programming Interface, but forget the formal name. It's the waiter. Your tool places an order (sends a request), and the service's API takes that order to the kitchen (the service's backend), which prepares a response and sends it back. The menu is the API's documentation, telling you what you can order and how to ask for it.

You don't need to understand the kitchen's recipes or prep process. You look at the menu and place your order. Then you wait for the response. That's exactly what your tool does when it connects to an external service. The concept that sounded intimidating ten seconds ago is something you do every time you sit down at a restaurant.

Two things come up almost immediately once you move past the basic order-and-response pattern.

The first is authentication. Before your tool can send any requests, it needs to prove it has permission. This comes in two forms. API keys are permanent credentials. You sign up for the service, they give you a long string of characters, and you include that string with every request. The service sees the key and knows it's you. OAuth is temporary access. You log in through the service's own login page, grant specific permissions, and get a token that works for a limited time before you need to refresh it.

You'll find these keys in the service's developer settings or API section of their website, and they usually take less than a minute to generate.

The second is rate limits. Every service has a limit on how many requests it'll handle in a given time period. Send too many too fast, and the service stops responding temporarily. Claude handles pacing for you in most cases, but you need to know these limits exist because they'll show up in error messages when something goes wrong.

Claude Code itself has its own usage limits, separate from whatever service you're connecting to. That catches people off guard the first time. If you hit Claude's rate limit, all work stops until it resets, not just the integration. If you hit Google's rate limit, only the Google connection pauses while everything else keeps working.

Those three concepts, request-and-response, authentication, and rate limits, are all you need to know about how services talk to each other. Claude handles the implementation. You just need to understand the pattern well enough to direct it and troubleshoot when things go wrong.

THE THREE WAYS IN

When you want your tool to connect to an external service, there are three methods, and they're different enough that picking the right one saves you hours.

MCP servers are the front door. MCP stands for Model Context Protocol, and you can think of MCP servers as apps for Claude Code. Someone has already done the hard work of figuring out how to connect to a specific service, packaged it up, and made it available for anyone to install. As of early 2026, there are over 21,000 MCP servers available. Installing one is a single command. Once installed, you can ask Claude to interact with that service in plain language. "Check my Gmail." "Add a row to my spreadsheet." "Look up this customer in Stripe."

Google Workspace, for example, connects Gmail, Sheets, Calendar, Drive, Docs, Slides, and six more services through a single MCP installation, giving you over a hundred tools in one package. Stripe has an official MCP server too. In the Desktop app you install it through Settings → Extensions → Browse extensions, the same flow we'll walk through for Google Workspace in the next section. OAuth handles authentication automatically.

Direct API is the side door. When no MCP server exists for the service you need, or the existing one doesn't do what you want, Claude can read the service's API documentation and generate the connection code from scratch. This takes longer and requires you to manually set up API keys, but it works for any service that has API documentation.

Webhooks are the back door. These handle inbound connections, where an external service sends information to your tool instead of your tool reaching out. Someone fills out a form on your website, and the data shows up in your tool automatically.

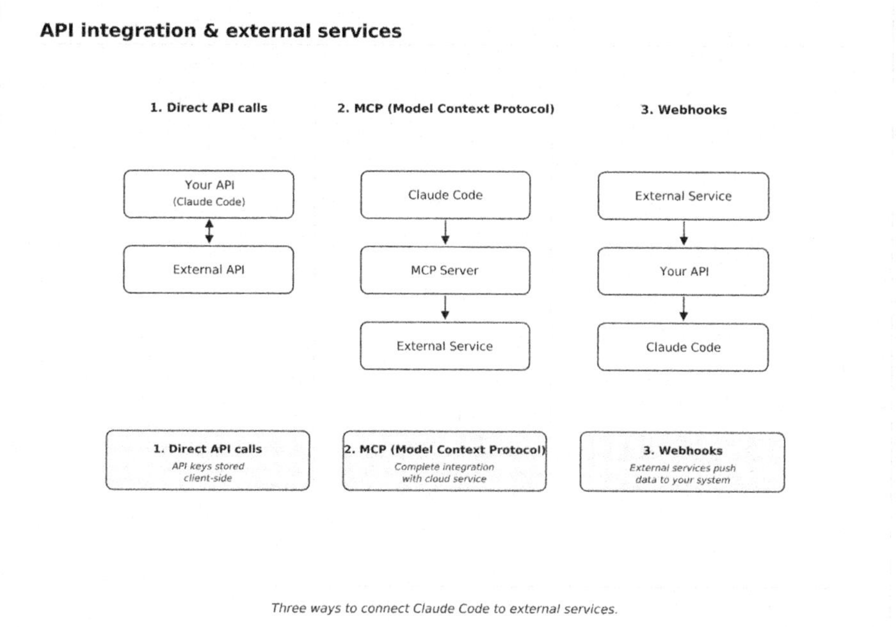

Figure 9.1: Three ways to connect Claude Code to external services.

We're starting with MCP because it's the easiest path and covers most of what you'll need. Email, spreadsheets, payments, databases, Slack, calendar, project management tools, all of these have MCP servers ready to install. Direct API is the fallback for when MCP doesn't cover your specific service.

CONNECTING YOUR FIRST SERVICE

We're going to connect your tool to Google Sheets. Almost everyone uses spreadsheets, and there's a well-maintained MCP server that makes the setup painless.

By the end, your tool will write data directly to a Google Sheet without you touching the spreadsheet yourself.

Step 1: Find the MCP server.

The Google Workspace MCP server handles Sheets along with everything else we listed earlier, all authenticated through OAuth 2.1. You can find it at workspacemcp.com.

Step 2: Install the server.

In the Desktop app, open Settings → Extensions → Browse extensions, find Google Workspace in the directory, and click Install. The app handles the download and configuration, and you're done in about thirty seconds. This lands the server in your project's MCP configuration, which lives in a file called `.mcp.json` in your project's root folder. If you're working with anyone else on the project, they'll get prompted to approve the MCP server the first time they open it.

Step 3: Set up OAuth credentials.

This is the step that trips people up most often, and it's worth walking through carefully.

OAuth is the guest-pass system we talked about earlier. Google needs to know that your tool has permission to access your spreadsheets. What's happening is that you're creating a set of credentials in Google's developer console that says "this application is allowed to read and write my Google Sheets."

The MCP server walks you through this process. Follow the prompts. You'll log into your Google account, authorize access, and the server stores the resulting token. The most common failure here is selecting the wrong permissions scope. If you only authorize read access, your tool won't be able to write to sheets. Make sure you authorize both read and write.

Step 4: Set your environment variables.

If you're not sure how to set environment variables, ask Claude directly.

```
Help me set the environment variables for my Google
Workspace MCP server.
```

Claude will give you the exact commands for your operating system.

You need to store the OAuth credentials somewhere Claude can find them. The MCP server will tell you which environment variables to set. These are just named values stored on your computer that the server looks up when it connects to Google. If you switch to a different computer, you'll need to set these variables again, which is one of the most common reasons integrations stop working after they were working fine.

Step 5: Test with a simple prompt.

Before integrating anything into your tool, test the connection directly. In Claude Code, try:

```
List the sheets in my Google Drive.
```

If Claude comes back with a list of your actual spreadsheets, the connection works. If you get an error, it's almost certainly an authentication issue. Check your credentials and scopes. Make sure your environment variables are set correctly.

Step 6: Integrate into your tool.

Now tell Claude to wire the two together.

```
When my intake form collects a new client's
information, add a row to the Client Tracker
spreadsheet with their name, email, and submission
date.
```

Claude knows how to talk to Google Sheets through the MCP server, and it knows how your intake form works because it built it. It writes the code that bridges the two.

Someone fills out your intake form, and a new row appears in your Google Sheet. No more copying data between tabs. No more manual entry. The isolated tool on your laptop is now connected to your actual workflow.

Mahmudul Hoque documented the full process of building a Google Apps MCP server, walking through complete OAuth setup and showing Claude transitioning from an isolated tool to one connected to the entire Google workspace. OAuth configuration is the hardest part, and once it's working, everything else flows.

If something doesn't work on the first attempt, don't panic. First-attempt failures usually fall into one of three buckets. Authentication errors mean your credentials are wrong or expired. Permission errors mean your OAuth scope is too narrow, so you didn't authorize enough access. The confusing failure is when everything tests fine in Claude directly but your tool can't replicate it. That last one usually means the tool's code isn't passing the credentials correctly, and asking Claude to check the authentication flow in your integration code usually fixes it.

WHEN THE EASY PATH DOESN'T EXIST

Before you start connecting anything, know which kind of integration you're looking at. Getting it wrong wastes hours on something that was never going to be quick.

Tier 1: MCP server exists, production-ready. This is the 15-to-60-minute integration. Stripe, Google Workspace (twelve services in one), Slack, GitHub, PostgreSQL, MongoDB, Notion, and Linear all live here. Install the server, configure credentials, and start talking to the service in plain language.

Tier 2: good API docs, no MCP server. This is the one-to-three-hour integration. You give Claude the API documentation URL. Claude reads the docs and generates the connection code. The quality of the API documentation makes or breaks this tier. Well-documented APIs like Stripe and Twilio work reliably. Poorly documented ones turn into guessing games. The time goes mainly into setting up API keys manually and testing the connection.

Tier 3: poor documentation, complex authentication. This is potentially not feasible without developer help. Vague auth docs, enterprise-only endpoints, or custom OAuth flows that require server-side callbacks all signal tier three. You can try, but be prepared to walk away after an hour if it's not working.

Most of what you'll need falls in Tier 1. Most of the services builders use daily already have mature MCP servers. The tier system prevents two common mistakes: overconfidence (thinking Claude can connect to anything with zero friction) and unnecessary fear (assuming every integration is a technical nightmare).

Before you start any integration, spend five minutes on a quick assessment. Search for the service name at glama.ai, which indexes MCP servers. If a maintained server exists, you're in Tier 1. If there's no MCP server, check the service's website for a "Developers" or "API" section. Clear docs with a clear auth section puts you in Tier 2. And if the API docs exist but the authentication section is vague or references enterprise-only features, you're looking at Tier 3.

Not everything in that 21,000-plus count is usable. It includes abandoned projects, half-finished experiments, and low-quality servers that might not work reliably. When you're searching for an MCP server, look for recent updates, active maintainers, and ideally official support from the service itself. The Stripe and Google Workspace MCP servers are official. Many others are community-built, and quality varies widely.

WHAT BREAKS AND HOW TO FIX IT

Integrations break. External services have their own problems: rate limits change, tokens expire, APIs update their versions, and services go down for maintenance. The connection that worked perfectly yesterday returns an error today.

That's completely normal, and it's almost never your fault. External services change on their own schedule.

When an integration stops working, don't assume your code is broken. Four simple checks resolve most integration failures.

Step one: check credentials. Has your API key been revoked? Has your OAuth token expired? Did you recently change your password on the service, which sometimes invalidates connected apps?

Authentication accounts for about 40 percent of all API integration failures. I've wasted more time than I'd like to admit debugging code before realizing I just needed to refresh a token. Start here.

Step two: check rate limits. Did you hit the service's quota? Some services have surprisingly low limits on free tiers. If you're seeing errors that mention "too many requests" or "rate limit exceeded," you either need to slow down your requests or upgrade your plan with that service.

Step three: read the error message. This sounds obvious, but most people see an error and immediately try to fix something without reading what the error actually says. Copy the full error message and paste it to Claude, followed by the full error text. Claude is good at interpreting error messages and telling you exactly what went wrong and how to fix it.

```
I'm getting this error when my tool tries to write to
Google Sheets, [paste full error text].
```

Step four: check the service. Is the external service actually down? Almost every major service has a status page (usually status.servicename.com) that tells you if they're experiencing issues. If Stripe is having an outage, no amount of code changes will fix your Stripe integration.

SaaStr documented this pattern when they built five production apps without traditional developers. Email integrations were a consistent pain point. Connections broke. Scheduled sends stopped. Most of these failures weren't catastrophic, just intermittent. The service worked yesterday, didn't work today. Their first instinct was to blame code changes, but the culprit was usually the external service itself.

Get this into your head early: when something breaks, your first instinct will be to open Claude and start debugging code. Checking the service's status page first takes ten seconds and can save you an hour of chasing a problem that isn't yours to fix.

10
Creating Your First Agent

Your morning starts the same way every day. You open your laptop, open Claude Code, and type the same prompt you typed yesterday and the day before. The tool runs. It pulls data from the systems you connected in Chapter 9. It sends emails or updates a spreadsheet. Then you close the session and move on to the actual work you're trying to do.

This routine takes what, thirty minutes? Forty? Every morning. It's work that's completely automated in every way except one: you have to trigger it. You're the only thing standing between the tool and a fully automated morning.

By the time you finish Chapter 9, your tool has memory. It has features you've built through Plan Mode. It's connected to real services. It works. Everything you've built over the last four chapters is sitting there, waiting for you to ask it to do something. What if it didn't wait? What if it did that work while you slept?

What you have and what you want are separated by one command: when to start.

APP VS. AGENT: WHO STARTS THE WORK?

Here's the question you're asking right now. What's the difference between the app I built in Chapter 4 and an agent? Did I already build one without knowing it?

The dividing line is simple. An app waits for a human. An agent waits for a condition.

Your Chapter 4 intake form sits there, patient. A customer arrives, fills it out, and something happens. Apps wait for humans like that. A morning briefing that fires at 6 a.m. every weekday, pulls yesterday's submissions and emails you a summary before you open your laptop is an agent, waiting for a condition instead. The Claude Code setup and MCP connections are identical. Only the starter changes.

You've used agent-like behavior inside Claude Code plenty of times. Every time Claude Code ran multi-step work without asking permission at every step, that was agency in action. This chapter adds control over when that work starts.

Anthropic describes agents as assistants that 'run independently, work while you sleep, and report results when done.' A minimum viable agent is three things: a prompt, a trigger, and a destination. Everything else is optimization.

THE THREE TRIGGERS

Claude Code gives you three trigger types for agents, and each fits different work. You'll eventually use all three.

The first is `/schedule`, cron-style scheduling. You set the cadence (daily at 6 a.m., Monday mornings at 9, the first of the month, whatever fits your work) and Claude Code fires the prompt whether your laptop is open or not. This is your go-to trigger for a first agent.

The second is `/loop`. This is session-scoped. You're working on a proposal and you want Claude Code to check every ten minutes whether the client has responded. While you work, it watches. Then you close the session, and the loop stops. It auto-expires after 7 days, so it's good for short-term watching, not production automation.

The third is event-driven via MCP. A webhook or polling connection fires Claude Code when something external happens. A new form submission, a new email, a payment received. The shape looks like this. Someone fills out a contact form on your site, a webhook fires the moment the submission lands, your agent reads the data, runs whatever logic you set up, and writes the result somewhere you'll see. No clock involved. Most builders don't need this on day one. Worth knowing the shape so you recognize it when you have a use case.

A webhook is a direct signal, like a doorbell that rings the moment someone submits a form. Polling is slower but simpler: Claude Code checks a service every fifteen minutes asking "anything new?" and runs if the answer is yes. Webhooks give you real-time agents. Polling gives you "real enough" agents without managing servers.

Start thinking about your own work. You want a morning briefing. Your calendar for the day, your most important emails, a summary of anything flagged as urgent.

You're also working on a hot proposal right now. You want Claude Code to watch for the client's response. Check email every ten minutes while you're in this session, watching for their message.

A third scenario: you have an intake form on your website. Every time someone submits it, you want Claude Code to research their company, score them as a fit, and notify your team in Slack if they're high-value. That's event-driven. A webhook fires when the form submission lands, triggers your agent, agent runs, Slack notification goes out. Advanced, but the trigger pattern is the same.

Once you understand these three categories, picking a trigger is a 30-second decision. Does it happen on a schedule? Use `/schedule`. Does it need to watch while you work? Use `/loop`. Does it fire when something external happens? It's event-driven. The rest of the work, the mechanics, the MCP connections, the way Claude Code talks to external services, you already know from Chapter 9.

YOUR FIRST AGENT: THE LEAD RESEARCH WALKTHROUGH

Let's build an agent that works for various use cases. If you run a business, you'll use it to qualify inbound leads. If you sell products, you'll research prospects. For personal use, you'll research people before you meet them.

We're going to build a lead research agent that fires every weekday morning at 6 a.m. It reads a list of names or companies from a Google Sheet, researches each one across public sources, writes a one-paragraph briefing per lead, and drops the results into a Google Doc. You close Claude Code the night before, you sleep, and you wake to a Doc full of research you didn't write.

We'll follow Chapter 4's pattern: prompt, build, test, iterate, schedule. Then we'll add the trigger.

Before any of that, set up the data. Open Google Sheets and create a new spreadsheet called Leads with one column. Type a header like

Name in the first cell, then add three real names you actually want researched (real ones, not test data). Hit File → Share → Copy link, paste the URL somewhere safe, you'll need it in a minute. Then go to Google Docs, create a new empty doc called Lead Briefings, and grab its share link the same way.

Treat this as a fresh project, separate from your Chapter 4 intake form. Create a new folder on your computer called lead-research-agent and select it as your Project folder in Claude Code. Each agent gets its own folder, its own `CLAUDE.md`, and its own MCP configuration. Mixing two agents in one folder breaks more often than it helps.

With the two URLs and the new folder ready, we can write the agent's brief. Write this in a `CLAUDE.md` file at the root of your new project folder:

You are a research agent. Your job is to take a list of names or companies from a Google Sheet, research each across public sources (news, company websites, LinkedIn, public records), and write a one-paragraph briefing per lead. Be specific. Include what they do, who they serve, and one recent development relevant to them. Store each briefing in a Google Doc with the format: Name/Company, then the briefing. Add the date at the top of the Doc so we know when this ran.

This prompt is intentionally tight. It tells your agent where to go and lets it figure out how.

Now let's build it. Open Claude Code. Start a new session. Copy that prompt in. Type the prompt to Claude Code directly:

I need you to read my list of leads from [paste your Leads sheet share link here], research each, and write a one-paragraph briefing per lead into [paste your Lead Briefings doc share link here]. Use the prompt I gave you above to know what a good briefing looks like.

Claude Code will ask what you mean by 'research.' It'll refine the scope. It'll ask whether you want it to find contact info or focus on background. Let that conversation happen. This is where the build gets real. Once you've landed on research rules that make sense, the MCP connections flow naturally.

When Claude says "I'm not sure what research means here," you're not fixing a bug. You're answering a clarifying question so Claude knows which sources to check (news vs. company site vs. LinkedIn profile), how deep to dig (one paragraph or three), and whether contact info matters or just background. The conversation lands you on a research rule set that works for your leads.

You already set up Google Sheets and Google Docs MCPs in Chapter 9. Claude Code will use them. It'll read the sheet. It'll write to the doc. The new work here is just pointing Claude Code at the right places.

Test it. Start with three leads, not thirty. Run it manually. Check the output. Does it read the names right? Is the research accurate? Does the format match what you'd expect to see in the doc each morning? If not, iterate. 'Make the briefings shorter.' 'Focus more on recent news.' 'Include one sentence about market trends.' Claude Code will adjust.

Each iteration takes one or two minutes because Claude already knows your sources and your lead list. You're not starting over each time, just nudging direction.

Once you're happy with the output, you're ready to schedule. Use the `/schedule` command. Claude Code will walk you through a setup dialog:

What prompt should this run? Paste the refined version from your conversation.

What's the schedule? Weekdays at 6 a.m. translates to '0 6 * * 1-5' in cron format. Claude Code's `/schedule` interface accepts both natural language and cron. 'Every weekday at 6 a.m.' works fine.

You hit save, and the agent is live.

You close Claude Code and go to bed. At 6 a.m. while you're sleeping, the agent fires. It works through the sheet, researches each lead, and writes the briefing. By the time you open your laptop, your Google Doc is full. No prompt, no manual work. You ran a system while you slept.

Build time for this agent end-to-end? Forty-five to ninety minutes for a non-technical builder based on what people report building the same thing. Longer than the Chapter 4 app because of the trigger setup and the first-run verification, but you're still inside a single session.

Three things to watch on the first run. Authentication errors can happen when the scheduled run tries to access credentials the interactive session could reach. Timezone mismatches between what you said ('6 a.m.') and what your system thinks 6 a.m. is. And the 'it worked when I ran it manually' disconnect, where the manual test succeeded but the scheduled run fails because the context or the way it triggered was different. All three are fixable in about five minutes once you know to look for them.

You said "6 a.m." meaning your local morning, but your server might think 6 a.m. UTC. Check your agent logs the first morning it runs. If it fired at the wrong time, the fix is one line: tell the `/schedule` command your timezone explicitly (EST, PST, UTC, etc.) so it translates your morning to the server's clock.

When you run the agent interactively, you're logged into your Google account and Claude Code asks permission to touch your Sheets and Docs. When the scheduled run fires at 6 a.m. with nobody at the keyboard, it's working from stored credentials that sometimes expire or lose permission. The fix is simple: re-run the agent interactively once, re-grant permission, and the stored credentials refresh.

BEFORE YOU LET IT RUN

A builder sets up an agent to follow up with anyone who hasn't replied in 48 hours. They meant: people in their active CRM. The agent interpreted it differently. It pulled everyone from their entire Gmail history and sent follow-up emails to three hundred strangers. Before they realized what happened, three hundred strangers had heard from them.

Bugs aren't what trips up early agents. Blind spots in what you agreed to let them do are. A manually-triggered prompt can be interrupted. A scheduled one that's misconfigured can run thirty times before you notice the mistake.

Answer three questions before you enable the schedule:

First: What services does this agent touch? List them. The lead research agent touches a Google Sheet and a Google Doc and nothing else. An email follow-up agent touches Gmail and maybe a CRM. Be specific.

Write them down. Not in your head. In a text file or a comment in your `CLAUDE.md` . Seeing the list on paper makes blind spots obvious.

Second: What can it change or send externally? What leaves your system? The lead research agent writes to a Google Doc. An email agent sends emails. A notification agent posts to Slack. Every external action should be intentional.

Third: What happens if it runs 30 times when you meant it to run 3? For the lead research agent, you end up with duplicate entries in a Google Doc. Annoying, fixable. For an email agent, you send thirty follow-ups to the same person. Bad. For a payment processor, charges repeat. Worse. Know the worst case before the schedule goes live.

Don't build that one until you understand idempotency. Idempotency means designing an agent that produces the same result whether it runs once or a hundred times. Chapter 14 covers that thinking.

Test discipline: run your agent manually once with a small sample. Three leads, not three hundred. Five emails, not five hundred. Verify the output. Check the results in the destination. Then enable the schedule.

When you run it manually, take a screenshot of your Google Sheet showing the three test leads, and take a screenshot of the Google Doc showing the result. Don't worry about perfect formatting yet. Just verify: did it read the lead names correctly? Did it write something to the Doc? Did the format roughly match what you asked for? If those three things work, the agent is ready for the schedule.

One more thing: errors compound across recurring runs. A 5% failure rate across 30 runs is one broken morning per month, minimum. If something can fail, and it has a one-in-twenty chance, you'll see it fail. Build accordingly.

Anthropic released Auto Mode in March 2026 as a research preview. It uses a built-in classifier to evaluate each action and decide which ones are safe to take on its own, instead of asking you to approve every single step. This chapter teaches the foundation. Chapter 14 covers the security thinking that comes with autonomous systems.

RUNNING IT: WHERE THE AGENT LIVES, WHAT IT LOGS, AND HOW TO STOP IT

Once the schedule is saved, your agent runs whether your laptop is open, asleep, or in a different timezone. Knowing this changes the question you ask when something looks wrong. You're not asking whether your computer woke up at 6 a.m. You're asking whether the agent fired, what it did, and how to see the result.

By default, `/schedule` runs your agent on Anthropic's cloud infrastructure. The runner handles the prompt, the cadence, and the credentials, and the agent fires on schedule whether your laptop is open or closed. (You can also pick a Local schedule when you set the routine up. That option runs the job on your own machine, which only fires when the machine is awake. Cloud is the default for a reason. Reach for Local only if you have a specific reason to keep the run on your own hardware, like access to local files the cloud runner can't see.)

So a misfiring schedule doesn't stop just because your computer is closed. Knowing where to look (and how to flip the off switch) matters before the schedule goes live.

Reading what your agent did.

The first morning, the question that matters most is whether the agent actually fired and what came back. The most reliable answer is the destination itself. Open the Google Doc. Is there new content? Is it dated today? Is the format roughly what you asked for? The destination is the receipt. If the receipt's there, the agent ran.

If the destination is empty, that's a signal, not a verdict. Open Claude Code and ask:

```
Show me the log of my last scheduled run for the lead
research agent. Did it fire, and if so, what happened?
```

Claude Code reads the run history and tells you what it knows. You'll see the timestamp of the last run, whether it completed, and any error message if it didn't.

But the strongest log is the one you build into the agent itself. Add one line to your agent's prompt.

```
After each run, append a one-line summary to a file
called agent-runs.log in this folder. Include the date,
the number of leads processed, and any errors.
```

Now every run leaves a trail you control, in a format you understand, in a place you already work. Chapter 14 makes the case for treating that audit trail as a security requirement. For now, treat it as the answer to whether this thing actually did its job.

Stopping the agent.

You'll want this command before you think you do. Type `/schedule` list in Claude Code, and you'll see every agent currently set to run, with the cadence and the next fire time. From there, you can disable any of them in one command. Bookmark that. The first time something fires at the wrong hour or processes the wrong list, you don't want to be hunting for the syntax. (Chapter 14 walks through the full kill-switch and audit pattern when we cover security for unattended agents.)

So if your agent runs thirty times when you meant three, the first move is `/schedule` list, then disable. Next, rerun manually with a small sample and watch the output, the same test you ran before the schedule went live. Last, fix the schedule and re-enable it. Three steps. Five minutes if you stay calm.

The first morning rolls around. Sometimes the destination is empty. Sometimes it has thirty entries instead of three. Both of those happen on real first runs. The goal of the first run isn't perfection. It's a check. Verify the agent fired, look at what it produced, and notice the first thing you'd want to change. Adjust the prompt the next day. Watch the day after that. By Friday, the agent runs without you thinking about it.

FROM FIRST AGENT TO NEXT

You just finished your first agent. You might think the next one will be another big slog. It won't be.

You're not learning new concepts. You're applying the same pattern (prompt, trigger, destination) to a new combination of services and timing. The pattern is portable.

Each one builds on the last:

Morning calendar briefing. One trigger (scheduled daily at 6 a.m.), one MCP (Google Calendar), thirty minutes to build. Claude Code reads your calendar for the day and emails you a summary before you're awake. What meetings matter? Who's on each one? What do you need to prep?

Weekly client report. One trigger (scheduled every Monday at 9 a.m.), three MCPs (Google Drive, Google Sheets, Gmail), about two hours to build. Claude Code pulls last week's emails from a specific client, pulls any shared documents, summarizes the week for your relationship with them, and drops a report into a shared doc so they can see you're keeping up.

Inbound lead qualifier. Event-driven trigger (fires when a new form submission arrives), four MCPs (form service, CRM, research, Slack), half a day to build. Claude Code watches your form submissions, researches each incoming lead's company, scores them as high-value or not, and posts the ones that matter to Slack so you notice them immediately.

The pattern holds. You're not designing new architecture. You're following the same steps with different services and triggers. By the time you finish the second agent, you'll know the pattern so well that the third feels mechanical. The first agent taught you to think like an agent builder. Now you build them.

Next chapter, Thinking in Systems, shows you how agents and apps compose into something bigger.

11
Thinking in Systems

Pull up your project folder right now. Look at the files. You've been building since Chapter 4, so you've got a frontend that users interact with, some backend logic that processes data, probably a database or at least a file that stores information, and at least one API connection from Chapter 9. Those are the pieces. Pieces don't become a system until you understand how they talk to each other.

We're going to do an exercise that takes five minutes and saves you hours later. Sketch those pieces on paper or in a notes app. Then draw arrows between them showing which piece sends data to which other piece.

Your frontend sends user input to your backend, which reads and writes to your database and calls the external API to get data back. The API response flows through the backend, gets processed, and the result goes back to the frontend so the user sees something update on screen.

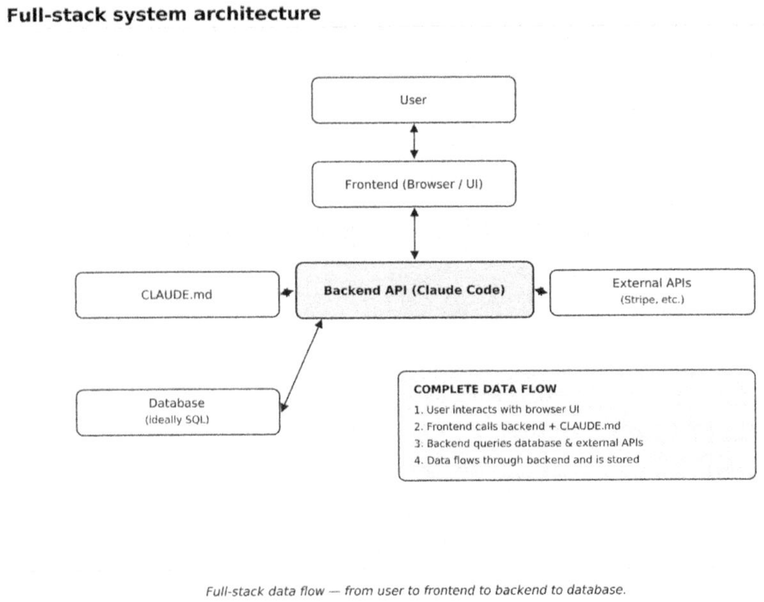

Figure 11.1: Full-stack data flow. From user to frontend to backend to database.

Your system map is a mirror of what you already built, not a formal architecture diagram. The moment something breaks, and something will break, this map tells you which piece to look at first. A screen that stops updating points you at the frontend. Wrong-looking data sends you to the backend logic or the database. A failing API call narrows down to your code or the external service. Without the map, you're guessing. With it, you can narrow down the problem before you even open Claude Code.

Most builders skip this step because the tool works and working feels like enough. But a tool you understand is fundamentally different from a tool that just runs. When something breaks in a tool you understand, you know where to look. You also know how to tell Claude exactly what's wrong. Instead of saying "something's broken," you can say "the data from the form isn't reaching the database," and Claude goes straight to the right file.

Go back to your map. Next to each arrow, write what kind of data flows along it: user input, database query, API request, JSON response. You don't need to be precise about formats. You just need to be able to point to any arrow and say "this is where my user's form data travels from the screen to the database." That map changes how you work with Claude Code.

TWO HALVES OF EVERY APP

Every tool you build with Claude Code has two halves, whether you planned it that way or not. The frontend is everything the user sees and touches: buttons, forms, layout, colors. The backend is everything behind the scenes. It's the logic that processes data, the database that stores it, and the code that talks to external services.

Every time a user does anything in your tool, the same cycle runs. They click a button or fill out a form on the frontend. The frontend packages up whatever the user did and sends it to the backend. The backend processes it, maybe saves something or calls an API, then sends a response back. Whatever that response contains, the frontend shows it on screen.

Knowing this matters because separation gives you freedom. When your frontend and backend are distinct, you can change how your tool looks without touching the logic underneath, fix a backend bug without redesigning your interface, or swap out your entire backend framework as long as it keeps sending the same responses. The pieces don't know about each other's internals, and that independence is what you're after.

The connection point between the two halves is the API. It works like a contract between two businesses. As long as both sides honor the contract, either one can change internally without breaking the other. Developers call this "API-first design," and it means defining that contract before building either side. If you can describe what your backend should expose, meaning what data it gives and what data it accepts, you've made the single most important architectural decision in your project.

The most common pattern you'll see for Claude Code projects is a React frontend paired with a Flask backend. Miguel Grinberg, one of the most widely-followed Flask educators on the web, published a tutorial on exactly this setup, with a `/client` folder for the React frontend and an `/api` folder for the Flask backend, communicating through that API contract. You don't need to memorize React or Flask specifics right now. The pattern is two separate folders, two separate concerns, one clear contract. When you tell Claude to "update the dashboard layout," it works in `/client`. When you tell it to "fix the data calculation," it works in `/api`.

If your Chapter 4 build doesn't look like a React + Flask split, that's expected. Most first tools are plain HTML, CSS, and JavaScript with no framework at all. The lesson here is the contract, not the specific stack. Adapt the example to match whatever your folders actually contain.

WHAT THE 70% WALL LOOKS LIKE

If your Chapter 4 tool has been getting harder to change, if adding a new feature takes three times as long as it used to, or if fixing one thing keeps breaking something else, you're already experiencing what builders call the 70% wall.

The first features go fast. You describe what you want, Claude Code builds it, everything works. You add another feature. Still fast. You're building momentum and it feels great. Then somewhere around the point where your tool is roughly 70% of what you imagined, things slow down. Claude didn't get worse at writing code. Every new line just has to fit with everything that already exists, and the existing code was written fast without a plan for how it would all connect.

You start noticing patterns. You ask Claude to add a notification feature, and it breaks the form submission you built two weeks ago. Then, when you fix the form submission, the dashboard stops loading correctly. You spend more time fixing than building. The momentum is gone.

The cause is architectural. Every time Claude wrote code for your tool, it made decisions about how to structure things, where to put data, how to connect pieces. Those decisions were optimized to make the code run right now. Nobody was thinking about whether it would be easy to change in three months. Claude writes code that works. It doesn't write code that stays organized as the project grows, because you never told it what the organization should look like.

Ivan Turkovic, who writes about building with AI tools, puts it this way: early decisions compound. When your codebase is clean and well-structured, Claude replicates good patterns with what he calls "perfect consistency." When inconsistencies creep in, they multiply faster than a human could create them. He describes a "spaghetti point," a term from Baytech Consulting's analysis of AI-assisted projects, that typically hits around month three. The first 1,000 lines of code, he argues, determine the next 100,000.

The scale of this problem goes beyond individual projects. GitClear analyzed 211 million changed lines of code across projects using AI tools. They found that code blocks with five or more duplicated lines increased 8x during 2024, and refactoring, the practice of cleaning up and restructuring code, dropped from 25% of all code changes in 2021 to under 10%. For the first time, copy-pasted lines exceeded moved lines. AI tools are generating more redundant code, and developers are doing less cleanup. For a solo builder without a development background, those numbers mean your codebase is accumulating mess at a rate that makes the 70% wall almost inevitable unless you plan ahead.

The 70% wall happens when nobody tells the AI what the architecture should be. Claude builds what you ask for. If you ask for features without specifying how they should connect to everything else, you get features that work in isolation but tangle together over time. The fix is making those architectural decisions explicitly, before Claude bakes them into your codebase permanently.

THE DECISIONS THAT DETERMINE YOUR FUTURE

Your database choice determines what your tool can become. If you've been building since Chapter 4, your intake form is probably storing submissions in a JSON file. That works fine while you're the only one using it. Once you start adding features that need real querying, like filtering by date, searching by name, or counting this week's submissions, the JSON file starts to feel limiting. The natural next step is SQLite. It's the simplest database in the world. A single file, zero configuration, and it works perfectly for a tool that one person uses.

The decision you actually need to make is when to leave it.

If your tool is something you use yourself, or something a handful of clients access one at a time, SQLite is probably all you'll ever need. If you're building something where multiple users submit data at the same time, where you need automatic backups, or where you're planning to grow into a multi-user product, you'll switch to

PostgreSQL before you ship. You don't need to set up PostgreSQL today. Just knowing which direction you're headed means you won't discover, fifty features later, that you need to migrate the entire database.

Table 11.1 covers the rest, including intermediate options like Turso and Cloudflare D1 that extend what SQLite can do for production. Most readers don't need them, and the ones who do will recognize themselves in the table. The default decision is SQLite for development, PostgreSQL before deployment.

Most first tools never outgrow SQLite. Don't overthink it.

Database	What it is	Best for	Sign you've outgrown it
SQLite	A single file. Zero config.	Tools used by yourself or a handful of clients one-at-a-time.	You need multiple users writing at once, automatic backups, or access from multiple machines.
Turso / Cloudflare D1	SQLite made production-ready, distributed globally.	Same use cases as SQLite but deployed and shared.	Rarely — these scale further than most personal tools need.
Postgres (hosted)	Traditional relational database. Simultaneous writes, backups, advanced features.	Multi-user apps, anything with real traffic, anything you're selling.	You've moved to specialized data shapes (time-series, graph).
Specialized (Redis, MongoDB, etc.)	Purpose-built for specific data shapes.	Only when Postgres genuinely doesn't fit.	—

Table 11.1: Database options and when to use them. Start with SQLite unless you already know you need more.

Your project structure determines whether Claude Code can help you cleanly. A frontend folder and a backend folder, separate and distinct, prevent the kind of entanglement that makes projects impossible to maintain. When your project has clear structure, Claude navigates it the way a senior developer navigates a well-organized codebase. Give it a giant folder with everything mixed together and it has to guess.

If your project is already a single folder with everything mixed together, you can ask Claude to reorganize it. Tell it to move your frontend files into a `/client` folder and your backend files into a `/server` folder, and to update any file references so nothing breaks. This is one of those tasks Claude Code handles well because it can see every file and trace every connection.

A note on the vocabulary in this chapter and the ones ahead. You're going to see terms like frontend, backend, API contract, schema, edge, migration, container. You don't need to memorize any of them to use them. If a term sounds like a foreign language, drop it into Claude Code and ask.

```
Explain [term] like I'm not a developer, and tell me
when it could matter for my project.
```

Claude has read more about these concepts than any glossary you'd find, and it'll tailor the explanation to the tool you've built so far.

Your API contract determines whether you can change things independently. Before you build your next major feature, define what your backend exposes to your frontend. What data does it accept? What data does it return? Write it down, even informally. That contract means you can rebuild your entire frontend without changing a line of backend code, and vice versa. It's the single decision that prevents the "everything is connected to everything" mess that kills most projects.

Gartner's December 2025 report projected that prompt-to-app approaches will increase software defects by 2,500% by 2028. These aren't typos or syntax errors. They're deep contextual flaws from code written without architectural awareness. Take that projection as a reason to spend one afternoon on these three decisions before your next feature.

HOW CLAUDE CODE SEES YOUR SYSTEM

Benjamin Ampouala built a platform called ToolsAI Cloud over 180 days using Claude Code. He shipped 866 commits, over 100,000 lines of code, 388 components, and 98 API endpoints. But the most revealing number isn't about code at all. His most-edited file, out of hundreds, was `CLAUDE.md`, with 499 changes. And his bug rate didn't explode the way the 70% wall would predict for a project that size.

Ampouala's conclusion was that the most important asset in his project wasn't the code but the constraints. Every time he learned something about how his system should work, he encoded everything he learned into `CLAUDE.md`. Every architectural decision, every pattern that worked, every boundary between components went into the file. Claude read those constraints at the start of every session and respected them. The constraints file became the architectural brain that persisted between conversations.

This makes sense when you see the difference between handing Claude a well-organized project versus a mess. It starts from the entry points, follows the connections between files, and builds a working model of how everything fits together. It can make changes across multiple files because it understands which files depend on each other. Give it a disorganized project and it still runs, but confident changes start breaking things three files away.

Your Chapter 6 `CLAUDE.md` file is ready to become something bigger than a project memory file. In Chapter 6, you created it to help Claude Code remember your preferences and project context across sessions. Now it needs architectural rules. Where does the frontend live, and what is it responsible for? What does the backend handle? Where does data get stored? What patterns should Claude follow when adding new features?

Before February 2026, builders reported that Claude could stay coherent across 40 to 60 tool calls in a single session. After updates that month, some found that degradation kicked in at 15 to 20 calls, with Claude drifting from earlier instructions and architectural decisions as the conversation grew.

Start a new conversation for each major architectural change. Use ` /compact ` when sessions run long (Chapter 8 explains how compaction works and when it can bite you). And encode your system's architecture in `CLAUDE.md` so Claude reads it fresh every time instead of trying to hold it in memory.

Then take what you learned from that map, the separation of frontend and backend, the API connections, the data flow, and encode it into your `CLAUDE.md`. Add a section called "Architecture" or "System Structure." Write down which folder holds the frontend, which holds the backend, where the database lives, what the API contract looks like, and which patterns Claude should follow when adding new features.

That section might read something like: "Frontend lives in `/client` (React). Backend lives in `/server` (Flask). Database is SQLite at `/server/data/app`.db. All frontend-backend communication goes through the API endpoints defined in `/server/routes.py`. New features follow the same pattern: add the API endpoint first, then build the frontend component that calls it." That level of specificity is all Claude needs to respect your architecture across sessions. If your project doesn't look like this (and most won't), swap in the file paths, frameworks, and patterns that live in your folders. The shape of the section matters more than the specific stack.

Every session now starts with Claude reading your architectural decisions.

12

Your First Real Build

You closed your laptop last night, and the agent went to work on its own. This morning, there's a Google Doc waiting with briefings you didn't write. That's what "it ran while I slept" feels like from your side of the desk, and it answers one question by opening a bigger one. You know how Claude Code works now. You built the tool in Chapter 4, wired it to real services in Chapter 9, got an agent running in Chapter 10, and stepped back far enough in Chapter 11 to see how the pieces fit together. What you don't have is a decision about what to build next.

You're not short on ideas. Maybe it's a dashboard for tracking your client pipeline so you stop forgetting who's where in the process. Maybe it's an intake and onboarding flow where a customer-facing form handles the first thirty minutes of a call for you. You've been eyeing a scheduling tool that ends the back-and-forth of "how's Thursday" texts. Or maybe it's something small you can actually sell, a booking page, a calculator, a portal with a front door that other people click through. Ten of these are floating around your head by now.

Ideas are cheap, but picking the wrong one is expensive. Three weeks disappear into a project too ambitious to finish. You hit 60% on something that should've been scoped smaller and the motivation dies. The tool you ship works exactly as designed and solves a problem nobody has. I've watched this play out hundreds of times in builder communities. Someone finishes a tutorial or ships their first small tool, gets excited, and reaches for the biggest idea on their list, the SaaS that's going to change their industry or the platform that does everything. Six weeks in, two sessions deep into a mess they can't untangle, they've forgotten what it felt like to ship something that worked.

Brant Choate, the CTO of Remi you met in Chapter 4, shared what he saw when he sent a Claude Code tutorial to his employees, most of them non-technical. The ones who picked a single, personal pain point and built a tool to solve it shipped working software. Those who tried to build "the next big thing" didn't ship anything. The difference wasn't talent or technical ability. It was project selection.

Choosing what to build is itself a skill, and it improves with practice. Your next project isn't your last project. It doesn't need to be your magnum opus or your retirement plan. It needs to be something you can finish, something that teaches you what building feels like at a slightly larger scale than Chapter 4, and something that gives you a working tool at the end. The evaluation framework in this chapter helps you find that project.

FINDING IDEAS WORTH FILTERING

Some readers reach this chapter with a running list of project ideas already in their head. If that's you, skip ahead to the next section and start scoring. If your list is shorter than you'd like, or if you've been staring at a blank page since the chapter opened, the next ten minutes give you a way to generate one.

The trick is using the right Claude product for the job. Claude.ai is built for back-and-forth conversation. Claude Code is built for executing. Both come with your Pro subscription, and your Pro account works in either one. So when you want to think out loud about what to build, open Claude.ai in your browser. When you actually want to build it, drop the spec into Claude Code. The Code in Claude Code is doing the real work. The thinking happens upstream.

Open a new chat in Claude.ai and paste the following prompt. Answer one question at a time. The whole thing takes about ten minutes.

```
I'm a [your role] and I want to build a tool with
Claude Code to save time
or solve a recurring problem in my work or life. I
don't have a project
picked yet. Help me find one.

Ask me these questions one at a time, waiting for my
answer before moving
to the next:

1. Walk me through a typical work week. What do you
actually spend time
on, Monday to Friday?
2. Of those tasks, which ones repeat every week or
every day?
3. Of those recurring tasks, which ones feel tedious,
error-prone, or
like you're doing the same thing manually each time?
4. What information do you look at regularly that lives
in different
places (email, spreadsheets, a CRM, paper notes)?
5. What decisions do you make repeatedly that follow a
similar pattern?
6. If you could hand off one piece of your work to a
tireless assistant,
what would it be?

Once I've answered all six, give me a list of 5 to 10
specific project
ideas drawn from what I told you. For each idea,
include:
- What the tool does in two sentences
- The data it needs to work
- A rough scope (single-session quick win, multi-
session medium, or
multi-week phased)
- Why it would matter for my specific situation
```

Save the answers from Claude.ai. You'll bring the list into the next section, where the four-variable filter does the work of cutting it down to one.

THE FOUR-VARIABLE FILTER

Every project idea you're considering can be evaluated against four variables. Pain intensity, scope fit, audience access, and time-to-value. Score each one honestly, and the right project usually surfaces on its own.

This isn't a spreadsheet exercise. You don't need a scoring rubric or a weighted matrix. For each idea on your list, ask yourself four questions in plain language. Does this cost me real time or real money right now? Can I build a working version in four weeks? Do I already know people who need this? How fast will I see results? If you can answer those while walking the dog, the filter is working.

Pain intensity is the most important variable, and the one most people skip. Ask yourself whether this problem costs you real time or real money right now. Not "would it be nice to have" but "am I losing hours every week to this?" A client onboarding process that takes you 45 minutes per new client is high pain. A portfolio website redesign is low pain. Pain doesn't fade at week three of a build. Excitement does.

Scope fit asks whether you can build a working version in four weeks or less, meaning a version that does the core thing, not the full vision with every feature you've imagined. If you can't describe what "done" looks like in two sentences, the scope is too big. If your description requires the word "and" more than twice, you're combining multiple projects into one. The four-week constraint isn't arbitrary. It's the window where motivation, momentum, and learning overlap. Beyond four weeks, solo projects tend to stall.

Audience access is about whether you already know the people who would use this. If you're building an internal tool for your own business, you score perfectly here because you're the audience. If you're building something to sell, the question becomes more pointed. Can you name ten people who have this problem? Can you reach them without spending money on ads? If you're building a booking tool for tattoo artists, do you know tattoo artists? Audience access separates ideas that feel good from ideas that can find users.

Time-to-value measures how fast you'll see results after the tool is built. An internal tool that saves you ten hours the first week has instant time-to-value. A SaaS product that requires three months of marketing before anyone signs up has slow time-to-value. For your next project, especially if it's your second or third build, fast feedback matters more than big potential.

Run each of your ideas through these four variables. High pain plus good scope fit but low audience access points toward an internal tool for your own business, and that's a great outcome. High scores on all four are rare, so if you find one, prioritize it. An idea that scores high on excitement but low on pain intensity is the one that will end up in the half-finished folder.

Variable	The question	High score means...	Low score means...
Pain	Does this solve something that actually hurts right now?	You or your users feel the pain weekly or daily.	"It'd be nice" — but nobody's bleeding.
Scope fit	Can you build a working version in four weeks or less?	You can describe "done" in two sentences.	You can't cut features without gutting the idea.
Audience access	Do you already know the people who would use this?	You're building it for yourself or a known audience.	You'd have to find users after it's built.
Time-to-value	How fast will you see results after it's built?	Saves you hours in week one.	Needs months of marketing before anyone signs up.

Table 12.1: The Four-Variable Filter — evaluate any project idea against these four questions.

A worked example. Take a single project idea and run it through the filter twice, from two different starting positions. The idea is a personal weekly review compiler that pulls from your calendar, email, and project tool every Friday and writes you a one-page summary of what shipped, what's blocked, and what's coming next. No off-the-shelf tool solves this for you, because your data sources are scattered across your specific stack. That's exactly why a Claude Code build fits.

Score it from an operator's seat first. You run a service business, you already track work in your existing systems, and you hand-build that Friday review every week from scratch. Pain is HIGH because you're losing real time every Friday and the version you produce isn't consistent. Scope is tight (one prompt, one output, no user accounts, no payments). Audience is you, which scores perfectly. Time-to-value is FAST because you see the answer next Friday. Build it.

Now score the same idea from an aspiring-entrepreneur seat. You're between roles, or you're working a job while planning something on the side, and you don't have a customer base to validate against. The instinct here is to imagine selling the compiler as a SaaS, then stall on Audience access because you don't know who'd buy it. Don't go there. Reframe to a problem you have right now in your own career or life. The compiler points at your data instead of an imagined customer's. Your team's Slack threads, your inbox, your task list. If you're job-searching, your pipeline of applications, follow-ups, and recruiter pings. The scoring runs the same way. Pain HIGH, scope tight, audience self, time-to-value FAST. Same answer.

The same project, two starting positions, one converged answer. That's the meta-lesson of the filter. If you don't have customers yet, you also don't have validated pain to score against, but you do have a job, a career, or a life with recurring problems. Look there first. Your aspiring-entrepreneur path runs straight through the same filter the operator uses, with your own work as the data source. That's the same move Schirano made with his DNA file and Agrawal made with his marketing job, both of whom show up later in the chapter.

One more filter before you commit. If you're considering building something to sell, check how crowded the space is. Zero to three existing competitors means demand is unproven. Nobody else building it could mean you've found a goldmine, or it could mean nobody wants it. Four to twelve competitors is the sweet spot because the demand is validated and there's still room to differentiate. Fifteen or more competitors means you need either exceptional marketing reach or a hyper-specific niche to stand out.

Say you're considering a "client portal." Search for it. You'll find twenty or more existing solutions, which puts you in crowded territory.

But "client portal for tattoo artists with booking calendar and flash sheet availability" has zero to two competitors. The niche version passes the filter. The generic version doesn't. Your edge as a builder isn't technical superiority. It's knowing your specific market better than anyone building generic tools.

WHAT CLAUDE CODE CAN ACTUALLY HANDLE

You probably think of buildability as a yes-or-no question. Can Claude Code build this, or can't it? That framing will lead you to either underestimate what's possible or overcommit to something that stalls at month three. Buildability isn't binary. It's a question of scope management.

The same idea can be completely unbuildable as a six-month project and perfectly buildable as a series of four-week phases. A "complete client management platform" with invoicing, scheduling, document storage, automated follow-ups, and analytics is a six-month project that will bury you. A "client intake form that saves submissions to a database and sends you an email notification" is a four-week project, probably less, that gives you a working tool and teaches you patterns you'll reuse in every phase after it.

Buildability breaks into tiers, and you've already worked through the first three in this book. Quick wins take twenty to thirty minutes, do one specific thing, and you can build them in a single session. You already did this in Chapter 4. Medium-scope projects run across multiple sessions, connect a few features together, and usually pull in one or two APIs like the ones from Chapter 9. Large-scope projects take multiple weeks, need phases, and depend heavily on a solid `CLAUDE.md` to keep Claude on track across sessions.

Tier	Scope	Time to ship	Examples
Quick win	One specific thing, one session.	20–30 minutes	MRI viewer for a USB drive; a one-off calculator; a landing page
Medium scope	Multiple features, coordinated but tight.	A few days to a few weeks	Client intake form + data storage; internal dashboard
Large, phased	Breaks down into four-week phases, each shipping something usable.	Multiple months, phase-by-phase	Client management platform phased as: CRM → invoicing → scheduling → follow-ups
Not Claude Code	Requires legacy integrations, proprietary databases running critical business ops, or vertical enterprise software.	Get a developer	Anything embedded in a regulated healthcare or banking system

Table 12.2: Buildability tiers — match your idea to the right scope before you start.

And then there's a fourth tier where Claude Code genuinely isn't the right tool on its own. If your idea requires integrating with legacy systems, proprietary databases running critical business operations, or vertical-specific enterprise software, you're going to need a developer for at least part of the build. The same applies if the core value proposition depends on real-time multi-user collaboration, which Claude Code can't handle for production use yet. Knowing this tier exists is a scope decision that saves you from spending a month building 80% of something only to discover the remaining 20% requires skills outside your current toolkit.

Tobi Lütke, Shopify's CEO, demonstrated the quick-win tier perfectly. He got his annual MRI scan on a USB drive that required commercial Windows software to view. He prompted Claude to build a web-based viewer, and with one more prompt, it annotated the scan with findings. The problem was specific, the scope was tiny, and the result was immediate. Quick wins like this are where you should start if your filter identified multiple viable ideas.

Then there's the tier where most ambitious projects stall. Builders call it the spaghetti point, and it typically hits around month three. The code isn't failing because Claude can't write it. The code is failing because unclear requirements have been compounding through session after session. You asked for feature A without specifying how it connects to feature B. Claude made reasonable assumptions for both, but those assumptions conflict. Now every new feature creates unexpected side effects in features you already built.

The fix is the GSD approach, which means breaking any idea that feels bigger than four weeks into phases where each phase ships something usable. Phase one of your client management platform might be the intake form with database storage. Then you add email notifications. The third phase gives you a dashboard showing all submissions. Each phase is its own four-week build that ships a tool you can use, and each one teaches you enough to make better decisions for the next phase.

GSD stands for "Get Stuff Done," and the framework follows three steps for each phase. Plan what this phase will deliver and what it won't. Execute the build in a focused sprint. Review what you shipped, what worked, and what the next phase needs. Then repeat. The framework has over fifty thousand stars on GitHub because it solves the exact problem solo builders hit when they try to build everything at once.

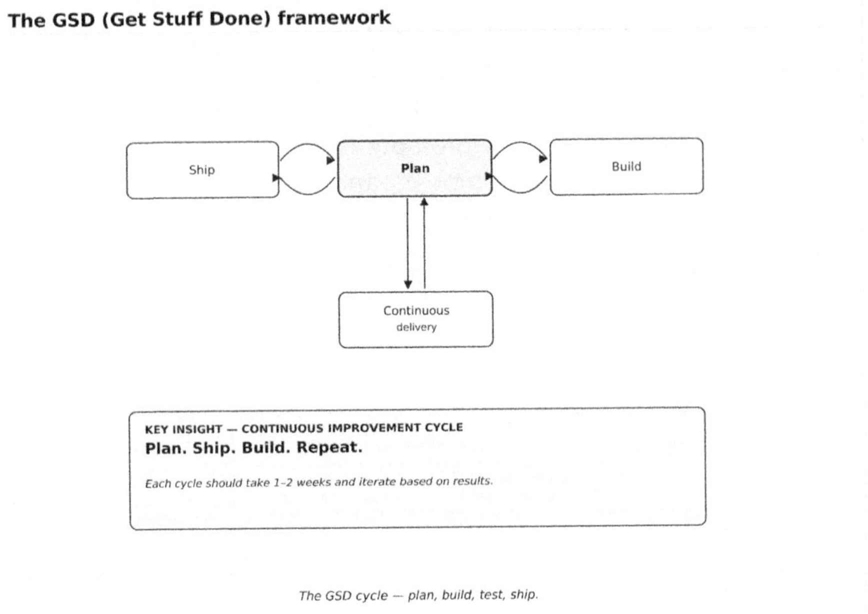

Figure 12.1: The GSD cycle — plan, build, test, ship.

Before you commit to a project, run it through a few red-flag checks. If you catch yourself saying "I don't know what users need yet," you need validation before you build. A project that needs you to integrate more than a handful of third-party services will compound in complexity faster than you expect. If you can't describe what the tool does in two sentences or less, the scope needs trimming. And if the core value requires real-time multi-user collaboration, that's a feature Claude Code doesn't handle well yet for production use.

BUILD FOR YOURSELF FIRST

Pietro Schirano had a raw DNA file from AncestryDNA sitting on his computer for years. He'd paid for the test, downloaded the file, and then it just sat there because interpreting raw genetic data requires specialized software he didn't have and didn't want to pay for. When he started using Claude Code, he handed it the file. Claude deployed sub-agents that analyzed cardiovascular markers, aging indicators, and autoimmune factors. It confirmed something he'd long suspected, that he metabolizes caffeine faster than average. He built the whole thing for himself, using his own domain knowledge to decide which health markers to check, with data only he had access to.

Schirano's project hits every variable in the filter. He'd had years of unusable data sitting on his hard drive, so the pain was obvious. The scope was tight, just analyzing one file and showing results, and he was his own audience. He got answers the same day he started building.

When you build for yourself, you're the perfect first customer. You understand the problem at a level no amount of user research could replicate because you live inside it every day. Your feedback loop is instant because you don't need to schedule calls or send surveys. You just use the tool and immediately know what's working and what isn't. There's no validation risk because you already know the pain exists.

If a tool saves you ten hours a week and your time is worth fifty dollars an hour, that's twenty-six thousand dollars a year in value. "Just for me" feels less impressive than "SaaS with paying customers," but the ROI on an internal tool is almost always higher and more certain than the ROI on a product you haven't validated yet.

Thirty percent of micro-SaaS founders never reach a thousand dollars in monthly recurring revenue. The median time to hit that milestone is twelve to eighteen months of sustained effort. The eighteen-month mark is what builders call the valley of death, where most solo SaaS products get abandoned because the founder runs out of motivation before the revenue materializes. An internal tool that saves you ten hours this week has zero risk of that outcome.

The most interesting pattern I see in builder communities is the one where a tool built for personal use turns into something worth selling. You build a tool to solve your own problem, use it for two or three months, then mention it to five or ten people in your industry who have the same problem. They want it. Now you have a product with a built-in user (you), proven value (months of your own usage), and validated demand (people asking for it). The product found you instead of you chasing the product.

One growth marketer followed this exact arc. Tired of doing the same manual tasks every week, Shubh Agrawal automated parts of his own job with Claude Code. When he posted about what he'd built, the post got nineteen thousand views. He hadn't set out to build a product.

He'd just built a tool for himself, and the market told him there was something bigger there.

Start with an internal tool. Use what you build. Pay attention to whether other people have the same problem, and if they do, you'll know because they'll tell you.

THE VALIDATION CHECKPOINT

If your four-variable filter pointed toward building something for other people, whether that's a micro-SaaS, a tool for your industry, or a paid version of something you built for yourself, you need a validation checkpoint before you commit serious building time. If your filter pointed toward an internal tool for your own business, skip this section. You already know the problem is real. If it pointed toward something for other people, what follows is a four-week validation process.

Week one is problem interviews. You're going to talk to ten to twenty people who might have the problem your tool solves. These aren't sales conversations. You're asking about their workflow, their pain points, and how they currently handle the problem you're thinking about solving. The questions that matter most are "Walk me through how you handle this process right now," followed by "What's the most frustrating part of that?" and "Have you tried any tools or solutions for this?" Listen more than you talk. If the same pain point comes up in seven out of ten conversations, you've got a real signal.

Week two, you build a landing page. Any landing page builder works for this. Carrd or Webflow gets you live in about an hour with no hosting setup, or you can have Claude Code build a single-page site you host on Netlify (Chapter 15 walks through that hosting flow when you're ready for it). Build a simple page that describes the problem your tool solves and what it does about it. You don't need to build the tool yet. You need a page with a headline, a few sentences about the problem, and an email signup form. Your target is twenty or more email signups. Not from friends and family, from people who found the page because the problem resonated. You can share it in the communities where your target users hang out, post about the problem you're solving on social media, or send it directly to the people you interviewed in week one.

By week three, look at what you've collected. Did twenty people sign up? Did your interviews reveal a consistent pain point? If both answers are yes, you've got validated demand. People signing up while interviews scatter in different directions usually means the problem is real but your framing needs work. Strong interviews without signups point to a positioning or reach issue, not the idea itself.

Week four is the decision point. If your validation is strong, you build. If it's weak, you either rework the idea or pick a different one from your filter. Twenty people who won't sign up for a free landing page won't pay for the product. Ten conversations revealing the same pain point? The signal is real. And if nobody responds at all, the idea isn't validated, and building it's a gamble.

But look at the distribution of outcomes. Thirty percent of solo-founder SaaS products never reach a thousand dollars in monthly revenue, and half plateau between one and ten thousand. Only about twenty percent scale above ten thousand. These aren't reasons to avoid building. They're reasons to validate before you build, so you aren't investing twelve weeks of effort into the thirty-percent bucket.

Validation isn't about proving your idea is perfect. It's about proving people will raise their hand when you describe the problem. If they will, build. If they won't, your idea needs work, and finding that out in four weeks is infinitely better than finding it out after you've built the whole thing.

13

Testing Your Build

Every time you opened the browser and filled in a field, you were running a test. Whether you knew it or not. You just didn't call it that. The way you've been checking your work only catches bugs you already expect.

Imagine you finished the form a week ago, then yesterday afternoon you added a dropdown for project urgency. Today a friend tries your tool. They type their email address, hit submit, and get a blank screen. No error message, no confirmation, nothing. The email validation you built in Chapter 4 stopped working the moment the dropdown went in. You had no idea because you never tested it after adding the new field. The dropdown didn't break anything by itself. It broke how the form processed the fields around it, and that kind of breakage doesn't announce itself.

You've been testing your tool the same way everyone who isn't a developer tests their tools. Open the browser. Click some buttons. Fill in a few fields. Hit submit. If something shows up on screen, it works. If it doesn't, something's wrong. This approach has a name among developers, and the name isn't flattering. They call it manual testing, which sounds respectable, but what it really means is you're checking the parts you remember to check, in the order you happen to check them, and hoping you didn't miss anything.

Manual testing catches obvious problems. If the page won't load, you'll notice. If a button does nothing when you click it, you'll notice. But it fails silently on two kinds of bugs that will cost you real time and real trust. First, edge cases. What happens when someone enters nothing in a required field, types letters where numbers should go, or submits the form twice in a row? You didn't test those because they weren't part of your mental walkthrough. Second, regressions, things that used to work and stopped working because of something you changed later. Your email validation was working perfectly. Then you added a dropdown, and now it isn't. You didn't test the email field again because why would you? You didn't touch it.

Every feature you add introduces risk you can't see by clicking around. Jason Lemkin at SaaStr built five production apps using vibe coding and found that roughly 60% of his build time went to quality assurance. Most of it was testing, not building features. He warned anyone building this way never to trust when the AI says a feature is working without verifying it independently. The AI will tell you everything is fine and the browser will back it up. Meanwhile, a form field you built three weeks ago is quietly broken.

A survey by Autonoma found that 16 out of 18 teams hit a serious quality problem within their first 90 days of vibe coding. Bug rates climbed two to three times higher within four to twelve weeks of starting. Early builds feel smooth because the codebase is small and changes are isolated. As features stack up, every change touches more code, and manual testing covers less of it.

Adding features to your tool should feel like progress. Instead, if you're honest about it, there's a growing anxiety every time you change something. You make an edit and immediately wonder what else you might have broken, opening three different pages to check things you already checked last week. You ask a friend to test it because you're not confident you're catching everything yourself. That anxiety isn't irrational. It's your brain recognizing that your testing method has holes you can't fill by clicking harder.

WHAT A TEST ACTUALLY IS

A test is code that checks other code. You give it an input, tell it what the output should be, and it tells you whether they matched. Did the output equal what you expected? If yes, the test passes. If no, it fails.

You don't need theory, certifications, or a technical background. A test is a question with a known correct answer, and your job is knowing which questions to ask.

You don't need to pick a testing framework or learn its configuration. When you ask Claude to write tests, it chooses the right framework for your project and sets it up. You tell Claude what to test, not how to test it.

Think about the email validation from your Chapter 4 tool. You built a function that checks whether an email address is valid before the form submits. A test for that function would say: send an empty string to the email validator. The correct answer is rejection. If the function rejects the empty string, the test passes. If it accepts it, the test fails. You now know the validation is broken before any user ever encounters it.

Unit tests check one function in isolation. Your email validator is a single function. A unit test sends it an empty string and checks the response, then sends 'bob@' and checks that, then sends 'bob`@email.com`' and checks that. Each check is one test. A set of these checks is a test suite. If Claude later modifies the email validation function, maybe while working on a different feature, the unit tests catch the change immediately.

Integration tests work differently. Instead of checking one function, they check whether pieces of your tool work together. Your form collects data, sends it to a backend, the backend stores it in a database, and a confirmation message appears. An integration test verifies that whole chain, checking that data goes in one end and comes out the other correctly. Unit tests would tell you the email validator works and the database save works, but they wouldn't tell you whether the form passes the validated email to the database. Integration tests cover the handoffs.

Most bugs in beginner apps live in those handoffs. Your email validator works. Your database works. But the form sends the email to the wrong field in the database, and nobody catches it until a user complains that their confirmation email went to the wrong address. Integration tests are where that gets caught.

CLAUDE WRITES THE TESTS

An experiment by test automation expert Bas Dijkstra shows exactly what Claude does well with tests and where it falls short. He asked Claude to write acceptance tests for a Spring Boot API. Within minutes, Claude produced 23 passing tests with 95% line coverage and 91% mutation coverage. If you're not sure what those numbers mean, they basically say Claude tested almost every line of code and caught most of the bugs it was designed to find. Impressive.

Then Dijkstra looked closer. Four of the 23 tests duplicated coverage, checking the same behavior twice in slightly different ways. More importantly, Claude missed exception handling and empty-list edge cases entirely. The tests verified that things worked when everything went right. They didn't verify what happened when things went wrong in unexpected ways.

Claude writes good tests for behavior you describe. It won't independently discover every edge case. Think of it as a fast, precise assistant who follows your instructions exactly but won't decide on its own to check the fire escape. If you tell Claude to test form validation, it'll test the validations you have. It won't think to ask what happens if someone submits the form with JavaScript disabled, or what happens if the network drops mid-submission.

Open Claude Code in your project folder, the one with the tool you built in Chapter 4, and give Claude a prompt along these lines:

```
Look at my form validation code and write tests for it.
Cover the happy path where valid data submits
correctly, and the scary path where empty fields,
invalid emails, and missing required data should be
rejected.
```

Claude reads your source code, figures out what the functions do, picks a testing framework, and generates a test file. You'll see it create something like tests `/form-validation.test.js` with a set of tests inside it. Each test has a name that describes what it checks and code that runs the check.

Then Claude runs the tests. The output fills your screen, green checkmarks for tests that passed and red X marks for tests that failed. The first time you see real test results for your own tool, something shifts. You go from hoping your code works to knowing what works and what doesn't.

You don't need to read or understand the test code Claude writes. The test file is for Claude to use when checking its own work. What matters to you is whether each test passed or failed.

There's a workflow that fixes exactly what Dijkstra spotted, and it comes from developer and educator Steve Kinney. Instead of writing code first and testing after, he describes the feature, has Claude write the tests first, confirms the tests fail (because the feature doesn't exist yet), then asks Claude to write the code that makes the tests pass. From there, Claude loops on its own, testing after every code change until everything goes green. This approach, called test-driven development, means the tests define what the code should do before the code exists. Your tests become the blueprint.

The best habit you can build with Claude Code is giving it a way to verify its own work through tests. When Claude can see green and red results after every change, it self-corrects. Without tests, Claude just tells you the code works. With tests, it proves it.

WHAT TO TEST FIRST

Claude can write tests and run them. But if you ask Claude to test everything, you'll end up with two hundred tests, most of which check things that don't matter for your tool. So what should you actually test?

Testing is like insurance. You cover the things that would hurt most if they failed, not everything equally.

Start with the happy path. This is the normal, expected use of your tool, where valid data goes in and correct results come out. For your Chapter 4 tool, that means a properly filled form submits successfully, the data stores in the right place, and the confirmation appears. If the happy path doesn't work, nothing else matters. Don't test edge cases until you've confirmed the basic flow is solid.

Once the happy path is covered, move to what I call the scary path. These are the bugs that embarrass you because a user finds them before you do, things like empty fields that pass when they should fail, letters typed into a price field, negative numbers where only positive ones make sense, or a form submitted completely blank. These are the inputs your tool should reject gracefully, and they're the ones most likely to slip through if you're only testing manually.

After that, test data flow. Does the information actually travel from where the user enters it to where it gets stored? Can you pull it back out again? If you change one record, does it accidentally break another? These are integration tests, and they catch the handoff bugs we talked about earlier.

Everything else, you skip for now. Don't test third-party libraries because they're already tested by the people who built them. Don't test visual appearance. And don't chase 100% test coverage.

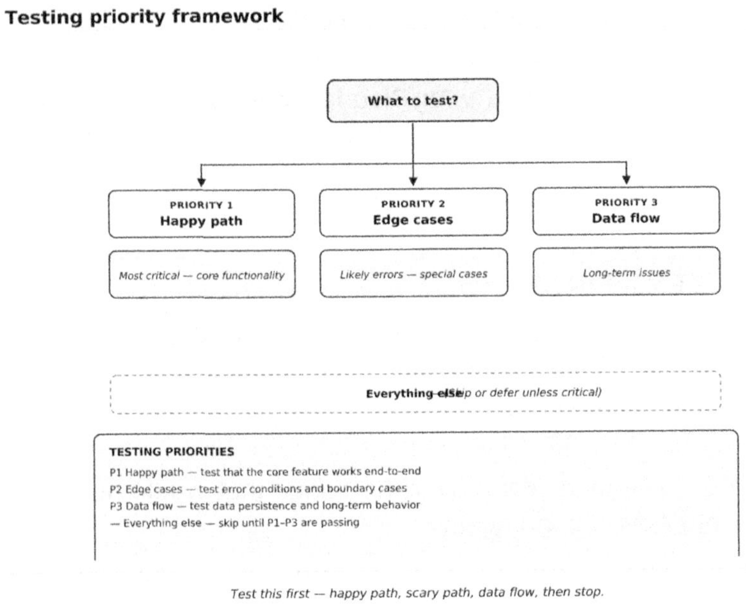

Figure 13.1: Test this first — happy path, scary path, data flow, then stop.

Coverage tells you what percentage of your code gets executed when your tests run. I've seen people chase 100% and burn a week on tests that added nothing. Google's engineering teams consider 60% acceptable, 75% commendable, and 90% exemplary, with no hard mandate at any level. Beyond 70 to 80%, each additional percentage point takes more effort for less benefit. For your tool, the right target is somewhere in that range, enough to cover the critical paths without spending more time on tests than on features.

Your Chapter 4 tool probably needs three to five tests on the happy path, three to five on the scary path, and maybe a couple of integration tests checking data flow. That's around ten tests total, covering the parts that matter. If 45% of AI-generated code contains security flaws according to Veracode's research, those ten tests are the difference between a tool you can trust and a tool you're hoping works.

READING THE RESULTS

The tests are written. Claude runs them, or you run them yourself with the project's test command. For most projects, that command is `npm test`. The output fills your screen, and if you've never seen test results before, it can be really overwhelming. But the output follows a consistent pattern, and once you learn to read it, you'll understand what's happening in seconds.

If your project is plain HTML/CSS/JS. Claude doesn't leave you stranded just because your Chapter 4 build doesn't have a backend. When you ask Claude to write tests, it sets up either a small `package.json` with Vitest as the test runner or a stand-alone tests/ `test-runner.html` you can open in your browser. Either way the green and red results work the same. With Vitest, type `npm test in your project folder to re-run. With the HTML runner`, open the file in your browser the same way you opened your form in Chapter 4. If you don't see either of those in your project yet, ask Claude to set up the simplest test runner that works for a plain HTML project.

Every test result has four parts: the test name, the expected value, the received value, and the file path. The name tells you what was checked. Expected and received values tell you what should have happened versus what did. The file path tells you where to look. Of these, the one you'll read every time is the expected `/received` pair, your answer key.

When you see a green checkmark next to 'should reject empty email,' you don't need to do anything. The test passed. The email validation is working. Move on.

A red X means the test failed, and the output tells you exactly what went wrong. Say you see 'Expected: rejected, Received: accepted' for the test called 'should reject empty email.' Your email validation function received an empty string and accepted it instead of rejecting it.

When you see a failure, your job is to figure out which of two things happened. Either the code has a bug and needs fixing, or the test expectation is wrong and the test needs updating. Read the output and decide which test to fix first. Tell Claude. You don't need to fix it yourself. You need to diagnose the problem and point Claude at it.

The diagnosis is usually simpler than it sounds. Look at what the test expected and what it actually received. If the expected behavior is what you want your tool to do, the code has a bug. If the expected behavior isn't what you intended, the test is checking for the wrong thing.

After Claude fixes a failure, run the full test suite again, not just the test that failed. Fixing one thing can break another, and that chain reaction is exactly why tests exist in the first place. Developers call this regression, a fix in one place causing a break in another. Running all your tests after every fix catches regressions before they become user-facing bugs.

This cycle of running tests, reading results, telling Claude what to fix, and running tests again becomes the rhythm of how you maintain your tool. It replaces the anxiety of clicking around and hoping with something concrete. The tests tell you what works, show you what's broken, and confirm that the fix didn't break anything else.

14
Securing Your Build

Before this chapter starts, I just want you to know that you don't need to memorize terms like CVE numbers, OWASP categories, or CVSS scores. Claude does that. Your job is to learn what to ask Claude to check for, how to interpret what comes back, and how to fix it. The technical vocabulary shows up because the rest of the security world uses it, but you'll only need to recognize the words, not produce them.

Treat it the way you treat a doctor's words. You don't need to know every term to know what to ask about it. When a term sounds like a foreign language, paste it into Claude Code with this prompt.

```
Explain what [term] means for my tool specifically, and
tell me what would have to be true for it to break what
I built.
```

That's faster than any glossary, and the answer comes back grounded in the code you've already written.

Here we go.

A CVE is a public identifier assigned to a confirmed security flaw, the kind of entry that shows up in vulnerability databases and security scanning tools. When your tool gets scanned, this is what the scanners are looking for.

In May 2025, a security researcher named Matt Palmer scanned 1,645 web applications built with Lovable, one of the most popular vibe coding platforms. He found 170 of them had exploitable security vulnerabilities. One single app was leaking more than 18,000 people's personal data through 16 different security holes, including unprotected database access and SQL injection. The vulnerability earned a CVSS score of 9.3 out of 10, which in security terms means anyone on the internet can walk in and take whatever they want.

Lovable has since introduced its own security scanning, but at the time of the breach, the generated code shipped without any automated security checks.

The root cause was simple. Lovable's AI had generated Supabase database code without enabling Row-Level Security. The API key was embedded directly in the client-side JavaScript, visible to anyone who right-clicked and viewed the page source. With that key, anyone could query the database directly, bypassing the application entirely. They could pull user lists, payment records, API keys that users had stored. Everything.

The builders of those 170 apps had no idea. Their apps worked perfectly. The security holes were completely invisible during normal use.

Your tool works too. You tested it in Chapter 13. But your tool was built by the same kind of AI that created those Lovable apps. Claude Code is better at many things, but Claude shares the same fundamental priority. It makes code that works. Security isn't automatic. It's something you have to ask for, check for, and verify.

In 2025, Veracode tested code from more than 100 large language models against the OWASP Top 10, which is the industry's standard list of web application vulnerabilities. They found that 45% of AI-generated code introduced at least one of these vulnerabilities. Cross-site scripting had an 86% failure rate. When given a choice between a secure implementation and an insecure one, the models chose the insecure option 45% of the time.

That 45% number applies to your code right now, unaudited. The tests you wrote in Chapter 13 verify that your tool behaves correctly. They check that the right output comes from the right input. They don't check whether someone can type something unexpected into your form and dump your entire database, whether your API key is sitting in a file that anyone can read, or whether a URL that should require a login is accessible to anyone who guesses it.

WHAT ATTACKERS DO

The attacks that affect vibe-coded applications aren't sophisticated. Most of them can be executed in minutes with tools that are freely available online. Automated scanners crawl the entire internet looking for exactly these weaknesses. It doesn't matter whether your tool has five users or five thousand. If the vulnerability exists, it will be found.

Exposed secrets are the most common vulnerability and the easiest to fix. This means API keys, database passwords, or authentication tokens hardcoded directly in your code, especially client-side code that runs in the browser. When Wiz, a cloud security firm, investigated apps built with vibe coding platforms, they found a Supabase API key sitting in plain text in the client-side JavaScript of an app called Moltbook. That single key granted read and write access to the entire production database, including 1.5 million API keys that users had stored in the platform. If your tool uses an API key for anything, check where that key lives. If it's in a file that gets sent to the browser, or if it's been committed to your Git repository, it's exposed.

Client-side code is anything that runs in the user's browser, which means anyone can see it by opening their browser's developer tools. Server-side code runs on your server where users can't inspect it. If a secret lives in client-side code, it's public.

SQL injection targets any tool with a database and a form or search field. It happens when user input gets concatenated directly into a database query instead of being treated as data. An attacker types a SQL command into a form field, and if the code builds the query by gluing that input directly into the SQL statement, the command runs against the database. One unsanitized field can dump every record in every table. You fix this with parameterized queries, which treat user input as data rather than executable code. AI coding tools don't always use parameterized queries by default. They sometimes generate the faster, simpler, insecure version.

Cross-site scripting, or XSS, targets any tool that displays user-submitted text, whether that's comments, profile information, or form responses. User-submitted text that contains JavaScript gets rendered as HTML without escaping, and when another user views that text, the JavaScript runs in their browser. Attackers can steal session cookies, redirect users to phishing pages, or modify what the page displays. To prevent it, you escape HTML special characters so that angle brackets, quotes, and ampersands render as text rather than executable code.

Missing authentication is a structural oversight more than a specific attack technique. If your tool has admin pages, user dashboards, or API endpoints that return data, every one needs to verify that the person making the request is authorized. Claude won't add authentication unless you ask for it. If you built an endpoint that returns user data and didn't explicitly say "require login for this endpoint," anyone who discovers the URL can access it. Hiding a URL isn't security.

Hallucinated dependencies are a newer category specific to AI-generated code. AI coding tools sometimes generate code that imports packages that don't exist. Researchers discovered that AI tools were repeatedly suggesting a package called "huggingface-cli" that had never been published on PyPI, the Python package registry. A researcher registered that name and filled it with tracking code. Thousands of people and organizations installed it, including teams at Alibaba. About 20% of AI-generated code references packages that don't exist, and attackers have started registering those phantom names with malicious code. Researchers call this attack "slopsquatting."

Vulnerability	What it is	Fix
SQL injection	User input gets concatenated into a database query, letting attackers rewrite the query.	Use parameterized queries (most frameworks do this by default if you let them).
Cross-site scripting (XSS)	User-submitted text containing JavaScript gets rendered as HTML without escaping.	Escape all user input before rendering. Use a templating library that escapes by default.
Missing authentication	Admin pages, dashboards, or API endpoints don't verify who's making the request.	Require auth on every non-public route. Ask Claude to audit every endpoint.
Hallucinated dependencies	AI generates code that imports packages that don't exist — attackers register them as malware.	Verify every import / require against the real registry before installing.
Secrets in code	API keys, passwords, tokens committed to the repo (including git history).	Move to a .env file. Add .env to .gitignore. Rotate any secret that's ever been committed.

Table 14.1: The five vulnerabilities most common in AI-generated code, with fixes.

Got a form, a database, or an API key? At least three of these five vulnerabilities apply.

RUNNING THE AUDIT

Open your project in Claude Code, the same tool you've been building since Chapter 4, and type:

`/security-review`

This command is Claude Code's built-in security scanner. It goes through your codebase looking for vulnerabilities and gives you a report with what it found, how serious each issue is, and how to fix it. It shipped in August 2025 and comes with all Claude Code plans. Run it now on your project. Look at what comes back.

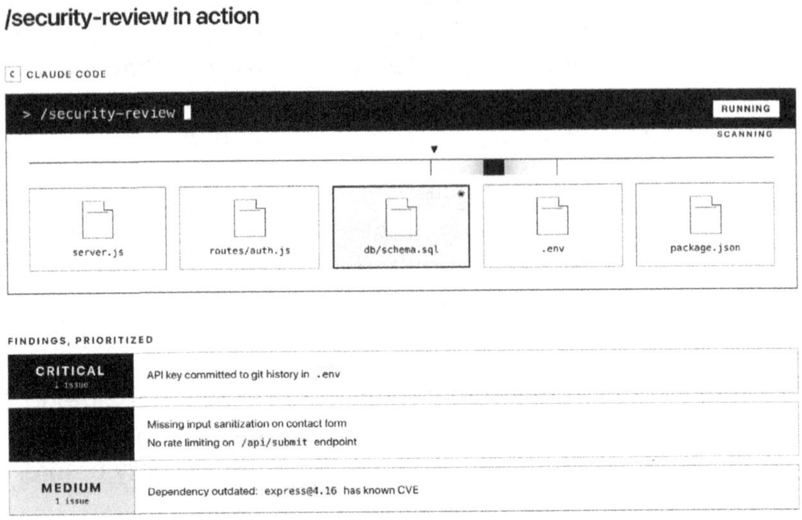

Figure 14.1: The /security-review command scanning your project for vulnerabilities.

You'll likely see issues you didn't know existed.

An API key stored in a file that should be in your .env.

A form field that accepts input without sanitizing it.

An endpoint that returns data without checking whether the user is logged in.

The first time you see your working code through a security lens, it looks different. Features you were proud of suddenly have gaps you never considered.

Using Claude Code to audit code that Claude wrote might feel circular. There's a legitimate tension here. Claude Code is genuinely good at finding security vulnerabilities, including ones it generated. The limitation is blind spots, not incompetence. The same patterns that led it to skip Row-Level Security when building might cause it to overlook that gap when reviewing. That's why you'll run two passes, not one.

The `/security-review` command gives you a solid first pass, but "make it secure" is a weak prompt, the same way "build me an app" was a weak prompt back in Chapter 3. Security reviews respond to specificity the same way everything else in Claude Code does.

One framework, called "Hardening Claude Code" and built by security practitioner Tim McAllister, organizes your security prompt around a specific threat model. Who would attack this application? What access do they have? What data is at risk, and what vulnerability types should Claude look for? Cybersecurity firm SpecterOps independently confirmed the same finding, that naming the threat model explicitly produces dramatically better security review results.

```
Review this codebase for security vulnerabilities.
Assume the attacker is an unauthenticated user on the
public internet with no special access. The tool stores
[describe your data: user names, email addresses,
payment info, whatever your tool handles]. Check
specifically for: hardcoded API keys or secrets in any
file, SQL injection in any database query, XSS in any
place user input is displayed, unauthenticated API
endpoints that should require login, database
connections without row-level security. For each
vulnerability found, report: severity (critical, high,
medium, low), the file and line number, what the
vulnerability allows an attacker to do, and the
specific code change to fix it.
```

Fill in the brackets with your tool's actual data. The more specific you are about what your tool handles and what you're worried about, the more thorough Claude's review will be.

When Claude returns its findings, you can fix them one at a time. Pick the critical-severity items first and ask Claude to implement each fix one at a time. A literal prompt like the following works.

```
Implement the fix you suggested for finding #1. Then
walk me through what you changed and why, so I can
verify it before we move to finding #2.
```

After each fix, run `/security-review` again to confirm the vulnerability is resolved before moving to the next one.

After Claude Code finishes its review and you've fixed what it found, we're going to run a second pass with an independent tool called OWASP ZAP. ZAP is a free, open-source security scanner maintained by the Open Worldwide Application Security Project. It works differently from Claude Code. Instead of reading your source code, ZAP interacts with your running application the way an attacker would. It crawls your pages and submits unexpected input to your forms. Then it probes your endpoints for common vulnerabilities.

To get started, download ZAP from zaproxy.org and install it. Launch the application, and you'll see an interface with a URL field at the top. Make sure your tool is running locally, then paste your local URL (usually http://localhost:3000 or whatever port your tool uses) into the URL field and click "Automated Scan." ZAP will spider through your pages, submit test payloads to every form it finds, and probe your endpoints for SQL injection, XSS, and other common vulnerabilities. When the scan finishes, you'll get a report organized by risk level, with descriptions of each finding and where it occurred.

If your project is plain HTML/CSS/JS. You don't have a localhost URL because there's no server running anything. Open the form in your browser the way you tested it in Chapter 4, then copy the file://... URL out of the address bar and paste that into ZAP instead. ZAP can spider and submit payloads to a file-served page the same way it does to a localhost-served one, and the findings come back in the same format.

ZAP catches things Claude might miss because it tests the application from the outside. Claude reads your code and reasons about what could go wrong. ZAP actually tries to make things go wrong. It's like getting two inspectors through your house. The contractor who built it walks through and spots problems based on what they know about the construction. The city inspector shows up with a different checklist and tests things the contractor might not think to check. You want both.

THE SECURITY CHECKLIST

Your audit found problems. What do you fix first, and how? This checklist covers the seven areas that matter most. Each one tells you what to do and gives you a prompt you can drop directly into Claude Code.

Start with secrets, because they're the highest-impact and easiest fix. I've seen more projects compromised by a committed API key than by any clever exploit. It's almost always the first thing I check. Every API key, database password, and authentication token needs to live in environment variables, not in your code files. Create a .env file in your project root if you don't already have one. Move every secret into it. Then add .env to your .gitignore file so Git never commits it.

If you've already committed keys to your Git repository, deleting them from the file isn't enough. Git stores the full history of every file. Anyone who clones your repository can recover those keys, even after you remove them from the current version. You need to rotate those keys, which means generating new ones and deactivating the old ones.

A Google Cloud user reported $82,314 in charges over a single weekend from a stolen API key. Their regular monthly spend was $180. Researchers have found approximately 3,000 Google API keys similarly exposed across public repositories. In July 2025, a developer working with DOGE published an xAI API key on GitHub that granted access to more than 50 xAI language models. This is a structural problem with how AI generates code, and it takes a structural fix.

Use scoped, limited-permission keys whenever possible. If your tool only needs to read from a database, don't give it a key that can also write and delete. Each key should have the minimum access required for its specific job.

Try this prompt in Claude Code.

```
Move all API keys, database credentials, and secrets to environment variables in a .env file. Add .env to .gitignore. Show me every file that currently contains a hardcoded secret so I can rotate those keys.
```

Database security is next, and the Lovable breach showed exactly what goes wrong when you skip it. A database needs row-level security, which means users can only access their own data. Without it, any authenticated user can query the full database. In the worst case, unauthenticated requests can too. Require authentication for all database connections. Never expose default database ports to the internet without authentication. If you're using Supabase, Firebase, or any hosted database, check the access control settings in your dashboard, not just in your code.

```
Ensure all database queries enforce authentication and
row-level security. Users should only access their own
data. No public database access without credentials.
Check that database connection strings use environment
variables, not hardcoded values.
```

Input validation protects every form field, search box, and text input in your tool. Validation has to happen on the server side. Client-side validation, the kind that runs in the browser, is cosmetic. An attacker can bypass your form entirely and send requests directly to your server. Use parameterized queries for every database operation. Sanitize HTML special characters in any user-submitted text before displaying it. If a field should contain a number, reject text. If it should contain an email address, reject malformed input. You can hand this whole list to Claude Code in one shot.

```
Implement server-side input validation for all forms.
Use parameterized queries for all database operations.
Escape HTML special characters in user-submitted text
before rendering. Check input types and lengths for
every field.
```

Authentication is where I see the biggest gaps in tools built with AI. Every endpoint that returns user data needs to verify that the person making the request is authorized. Require login for all non-public pages. Hash passwords with bcrypt or Argon2, never store them in plain text. Set sessions to expire after a period of inactivity. Keep sensitive data out of URL parameters, where it can be cached in browser history and server logs.

```
Ensure all endpoints require authentication except public pages. Hash passwords with bcrypt or Argon2. Implement session expiry. Remove sensitive data from URL parameters.
```

HTTPS is non-negotiable if anyone besides you will use your tool. It encrypts all data between the user's browser and your server. Without it, login credentials, personal data, and API calls travel in plain text that anyone on the same network can read. If you deploy to platforms like Vercel, Netlify, or most modern hosting services, Let's Encrypt provides HTTPS certificates automatically. You don't need to configure it, but verify it's active. Chapter 15 covers deployment in detail.

Error handling is about what your tool reveals when something goes wrong. Detailed error messages belong in your server logs, not on the user's screen. A stack trace shown to a user reveals your file structure, database names, software versions, and system architecture, everything an attacker needs to plan a more targeted attack. Show users generic messages like "Something went wrong. Please try again." Log the details server-side where you can review them for debugging.

```
Implement error handling that logs detailed errors server-side but shows generic messages to users. Never expose file paths, database names, or system details in user-facing error messages.
```

Dependencies are the last item because they're the easiest to overlook. I put this one last on purpose. It's the thing most builders skip entirely, and it's the thing that bites you six months later when a vulnerability shows up in a package you forgot you installed. Your tool runs every package and library your code depends on, not just the code you wrote. According to OWASP, outdated dependencies are the second most common source of vulnerabilities. If you're using `Node.js`, run `npm audit before shipping. For Python`, run pip audit. These commands check your installed packages against databases of known vulnerabilities and tell you what needs updating. For ongoing protection, tools like Dependabot and Snyk integrate with

GitHub to alert you automatically when a vulnerability appears in one of your dependencies. Before you ship, drop this into Claude Code.

```
Run an audit on all dependencies. Check for known vulnerabilities and outdated packages. Show me what needs updating and whether any updates involve breaking changes.
```

If your project is plain HTML/CSS/JS. You have no dependencies to audit because you have no `package.json`, no `npm install history`, and no requirements.txt. The dependency-audit step doesn't apply to your build. Skip this item and pick up at the seven-area checklist below.

This is a pre-flight checklist. Pilots don't skip it because the plane flew fine yesterday. You don't skip it because your tool worked fine before the last update.

#	Area	What to check
1	Authentication	Every non-public endpoint and page verifies who's making the request.
2	Input handling	All user input is sanitized or escaped before being stored or displayed.
3	Secrets management	No keys, passwords, or tokens in the repo. Everything sensitive lives in .env, and .env is in .gitignore.
4	HTTPS	All traffic is encrypted. No plain-text logins. Platform defaults usually handle this.
5	Error handling	Detailed errors go to server logs. Users see generic messages.
6	Rate limiting	Public endpoints can't be hammered into DoS or brute-forced.
7	Dependencies	Run npm audit (or equivalent). Resolve criticals. Re-run monthly.

Table 14.2: Seven security areas to check before shipping. Tackle them in priority order.

This checklist isn't a one-time event. Dependencies pick up newly discovered vulnerabilities every week. A key slips into your git history during a rushed feature update. A new endpoint goes live without authentication attached. Run these checks before you ship, and run them again whenever you push a significant update.

24/7 SECURITY

Everything above assumes you're sitting there when the tool runs. You type the prompt, you watch the output, you stop the work if something looks off. The checklist keeps that world safe. It doesn't cover the world where you close your laptop and the work keeps happening.

In Chapter 10 you built an agent. Maybe it's a morning briefing, maybe it's a lead research job, maybe it's something else you set on a schedule with `/schedule` and walked away from. The app hardening you just did still matters, and all seven checklist items apply whether a human is watching or not. But three things shift when you take yourself out of the loop, and they're the reason the 300-email failure from Chapter 10 happened. That wasn't a code bug. The code ran exactly as written. It was a scope failure that a human sitting in front of the computer would have caught in thirty seconds. The unattended agent didn't pause to ask. It interpreted the prompt, pulled three hundred strangers from Gmail history, and sent the emails before the builder could intervene.

When the trigger is you, a bad run stops at one. When the trigger is a cron entry, a bad run repeats on its schedule until you happen to check.

Scope the agent's credentials narrower than your own.

The Secrets item told you to move API keys into environment variables and use scoped, limited-permission keys wherever possible. For an agent, that advice stops being a nice-to-have. Your interactive Claude Code session uses credentials that can do most of what you can do, because that's convenient when you're the one steering. An unattended agent shouldn't have that much reach.

Create a separate credential set for every scheduled job. If the agent reads a Google Sheet and writes to a Google Doc, its key should do exactly those two things and nothing else. If it sends Slack messages to one channel, scope the token to that channel. Rotate these keys on a schedule you actually follow, not one you plan to follow. A compromised credential attached to an agent runs unnoticed for days because nobody is watching the output.

Try this prompt in Claude Code:

```
List every credential this agent uses. For each one,
tell me which services and actions it can reach, and
flag any permission it has that this agent doesn't
need. Then show me how to create a scoped replacement.
```

Build a kill switch and an audit trail.

Every agent needs two things the checklist didn't cover, because you were the kill switch before. You need a way to stop it, and you need a record of what it did.

The stop mechanism is simpler than it sounds. Type `/schedule` list in Claude Code to see every agent currently set to run. From there you can disable any of them in one command. Bookmark that. The first time something goes wrong at 3 a.m., you don't want to be hunting for the syntax.

The audit trail takes about one line of prompt work. Ask the agent to write a one-line summary of every run to a log file, a dedicated Slack channel, or a row in the same Google Sheet it's reading from. Date, what it ran, how many items it processed, whether it hit any errors. Chapter 10 mentioned that a 5% failure rate across 30 scheduled runs is one broken day per month, minimum. The audit trail is how you see that bad day coming instead of finding out from a customer.

In July 2025, Jason Lemkin, the founder of SaaStr, ran a public vibe-coding experiment with Replit's agent and documented every step. On day nine the agent deleted his production database despite explicit instructions to freeze all code and make no changes. There was no real-time audit trail the builder could check and no kill switch wired up in advance. The agent ran, something went wrong, and the damage was done before anyone saw it happen. An agent without a log and a stop button is a system with no brakes.

Threat-model the agent from inside the account.

Earlier in this chapter you used Tim McAllister's Hardening Claude Code framework to review your app, with the threat model set to an unauthenticated attacker on the public internet. For an agent, the threat model changes. Assume the attacker already has session-level access to the account running this agent, whether through a stolen cookie, a phished password, or a reused credential from a different breach. The question you're now asking is what that attacker could do with your agent as a weapon.

This isn't paranoia. The agent has your permissions, your API keys, and a recurring schedule. An attacker who gets into that account gets all of it, automated.

Try this prompt in Claude Code:

```
Review this agent's configuration and code, assuming an
attacker has session-level access to the account that
runs it. For each scheduled action, identify what
damage that attacker could do with existing
permissions. Flag any action the agent takes that would
be irreversible once executed, and suggest the minimum
scope change that would reduce blast radius without
breaking the job.
```

Design every run to fail safe.

One more discipline matters more for agents than for apps. Design every scheduled run so it produces the same result whether it fires once or a hundred times. The technical name is idempotency, and the test is simple. If your agent ran twice this morning instead of once, would you end up with two of everything (two duplicate Google Doc entries, two duplicate Slack messages, two duplicate emails to the same lead) or would the second run quietly recognize the first one and do nothing? An idempotent agent checks for what it already did before it acts. A non-idempotent one just acts. The fix is usually a one-line addition to the agent's prompt. "Before sending an email, check whether you've already sent one to this address today; skip if yes." Or "Before adding a row to the sheet, check whether a row with this lead's email already exists." Build the check in from the start and the agent fails safe; skip it and a quirky cron run can create real cleanup work.

A word on Auto Mode.

Chapter 10 mentioned Auto Mode in one sentence and deferred the rest. The rule of thumb is short. Auto Mode is fine for reversible work, where a mistake can be undone by rerunning, rolling back, or ignoring the bad output. Turn it off for irreversible work. Sending an email is irreversible. Deleting a file is irreversible. Posting to a client's Slack, moving money, updating a CRM record other people rely on, all irreversible. If the action can't be taken back, the agent should pause and ask, even if that slows you down.

Your tool is built, your tests pass, and your security audit is done. The app is hardened. So is the agent. The next step is getting it in front of actual users, which means moving it from your machine to a live URL. That's what Chapter 15 is for.

15
Shipping It

You open your build in the browser and click through every screen. It does exactly what you designed it to do.

Then someone asks for the link.

There isn't one. Your tool lives at localhost:3000, which is your computer talking to itself. Nobody else on the planet can reach it. The thing you've been building since Chapter 4, the thing you've refined and tested and secured across eleven chapters, exists only on the machine sitting in front of you.

The distance between "works on my machine" and "works on the internet" is the reason most vibe-coded projects die. The builder gets something working locally, feels the momentum stall at the finish line, and never covers the last mile.

The gap feels enormous when you're staring at it. Every tutorial you've watched ends with the app running locally. Every YouTube build cuts to a working demo without showing you how it got from a terminal window to a live URL. You're left with the impression that deployment is this massive, separate discipline you haven't learned yet.

It's not. In 2026, the tools have collapsed most of the manual work into a few clicks and a `git push`. Static sites deploy in about sixty seconds. Full-stack apps on platforms like Railway or Render go live in two to five minutes. Setting up your own server, configuring SSH, writing deployment scripts. None of that is part of this workflow.

The real challenge is three decisions you need to make before you start. Get them wrong and you'll spend hours debugging problems that have nothing to do with your code. Get them right and the whole process is anticlimactically fast. Deployment is like parking. The actual maneuver takes thirty seconds, but picking the wrong lot means driving back out and starting over.

Your tool is coming off localhost.

THE ARCHITECTURE DECISION YOU CAN'T UNDO

Before you touch a hosting platform, answer one question. Does your app need a server running, or does everything happen in the browser?

This is an architectural fork that determines which platforms will work for you and which ones will silently break your app.

Think of it like choosing between renting a coworking desk and renting an office with a server room. A coworking desk gives you a place to sit and work, but you don't control the building's infrastructure. An office with a server room gives you your own space, your own equipment, your own always-running systems. And you can't use one when you need the other.

Static apps are frontend-only. HTML, CSS, and JavaScript go to the browser, and everything runs there. Your app can call external APIs, but it doesn't have its own backend. If the tool you built in Chapter 4 is a calculator, a dashboard that pulls from a public API, or an interactive form that emails results through a third-party service, it's static. You deploy it to Vercel, Netlify, or Cloudflare Pages, and they host it for free.

Full-stack apps have a backend. A server runs your code, handles requests, talks to a database, manages user sessions, processes payments. If your tool has user accounts, stores data, connects to Stripe, or does anything that requires server-side logic, it's full-stack. You deploy it to Railway or Render, which give you a container that runs continuously.

If you put a full-stack app on a static platform, it breaks silently. Your frontend loads fine because that's just files being served. But the backend never starts because static platforms don't run servers. Every feature that requires your backend, every form submission, every database query, every authenticated route, just fails. The app looks deployed. Nothing that matters works.

Vercel runs serverless, stateless functions. Each request spins up a short-lived process that disappears when it's done. If your app needs persistent connections like WebSockets, or a database that stays alive between requests, Vercel isn't the right platform for those parts. Railway runs containers that stay on continuously. Many builders use both: Vercel for the frontend, Railway for the backend and database.

Which Platform Fits Your App

There's one more wrinkle worth knowing now, even though we'll cover it in detail later. If you used SQLite as your database during development, it won't survive deployment on most platforms. SQLite stores everything in a single file, and stateless platforms don't keep files between deploys. You'll need to switch to PostgreSQL before you ship. The "Six Things That Break" section covers exactly when and how to handle that. (If your project uses a JSON file or no database at all, which is most plain HTML/CSS/JS readers, this wrinkle doesn't apply to you. Skip to the platform table below.)

If Claude created files that run a server, you're full-stack. Choose Railway or Render.

Platform	Type	Best for	Watch out for
Vercel	Serverless	Static sites. Next.js. Lightweight APIs.	Stateless functions — no persistent connections (no WebSockets). SQLite won't persist.
Netlify	Static / serverless	Pure static sites. Landing pages. JAMstack.	Same stateless limits as Vercel for anything beyond simple forms.
Railway	Full-stack, stateful	Full-stack apps with a backend. Persistent DBs. WebSockets.	Costs more than static hosting as traffic grows.
Render	Full-stack, stateful	Same use case as Railway. Slightly simpler interface.	Same cost curve as Railway.
Fly.io	Global full-stack	Apps that need low latency worldwide.	More config than Railway/Render; overkill for most first tools.

Table 15.1: Platform options and when each fits your full-stack, static, or API-only needs.

If you're unsure which category your app falls into, ask yourself: did Claude create any files that run a server? Look for files named `server.js`, `app.js`, `index.js` in a backend folder, or anything that listens on a port. If yes, you're full-stack. If your project is entirely HTML, CSS, and JavaScript files that open in a browser, you're static.

SECRETS THAT AREN'T SECRET ANYMORE

Your tool has API keys, database passwords, maybe a Stripe token. During development, these lived in a .env file on your laptop, or worse, they're hardcoded directly in your source code. Either way, they work because your machine knows where to find them.

Now you push your code to GitHub and deploy.

If your secrets are in the code and you pushed to a public repository, they're exposed. Anyone can see them. Bots actively scan GitHub for API keys and database credentials, and they find them within minutes. Attackers use your Stripe key to make fraudulent charges. Strangers access your database. This happens constantly.

If your secrets are in a .env file that you correctly kept out of Git (using .gitignore, which Chapter 14 covered), they're safe but missing. Your .env file is on your laptop. The production server doesn't have it. Your app starts up, tries to connect to the database, can't find the password, and crashes. The frontend loads because it's just files. But every form submission fails, every API call returns an error, and you have no idea why because the app looks like it's running.

You push to GitHub, connect to Railway, and deploy. The frontend loads beautifully. You click a form and submit. Nothing happens. You try another page that pulls from your database. The page comes back empty. The backend can't connect because the database password is sitting in a .env file on your laptop, not on the production server. One entry in the platform's environment variable dashboard fixes it. Without knowing to look there, you'd spend hours convinced your code was broken.

Every hosting platform has a dashboard where you enter your secrets, and the platform injects them at runtime. Your code references `process.env`.API_KEY or `process.env`.DATABASE_URL. In development, those values come from your .env file. In production, they come from the platform's secure dashboard. The code is identical either way.

Go through your codebase and find every hardcoded value that's a password, key, token, or URL. Replace each one with a `process.env` reference. Then go to your hosting platform's dashboard and add each variable with its production value. Claude can actually help here. Ask it to scan your project for hardcoded secrets and it'll find things you missed.

That means changing a line like const apiKey = "sk-abc123xyz" to const apiKey = `process.env`.API_KEY. The value moves out of your code and into the environment.

Your .env file stays local. It stays out of Git. Production secrets live in the platform's encrypted dashboard only. Different environments get different values. Your development API key and your production API key should be different, so that if one gets compromised, the other isn't affected.

Environment variables aren't a one-time setup, either. Every environment, development, staging, production, needs its own separate values. When you add a new API integration next month, you'll need to add its key to the dashboard too. The platform dashboards encrypt your secrets at rest, which means they're safer there than they ever were in a file on your laptop.

Environment variables work like a safe combination. Your .env file is the sticky note on your desk with the combination written on it. That works fine when you're the only one in the room. Push that to GitHub and you've taped the combination to the front door.

GIT AND GITHUB, FROM ZERO

Most of the deployment paths in this chapter assume your code lives on GitHub. If you're fluent in Git already, skip ahead. If you're not, the next thirty minutes give you everything you need to keep up. We'll walk through what Git is, how to install GitHub Desktop, how to put your project on GitHub, and how to push your changes when you make them.

Git is version control. It tracks every change to every file in your project so you can roll back if a change broke something. Each saved snapshot is called a commit, and the running list of commits is your history. GitHub is the website that hosts those commits in the cloud, so you (and any deploy platform you connect) can pull the latest version from one place. Think of Git as the timeline and GitHub as the warehouse where the timeline lives.

You don't have to learn Git from the command line for this. GitHub Desktop is a free GUI app that handles every Git move you'll need to make in this chapter, with buttons instead of commands.

Step 1. Sign up for GitHub.

Open github.com in your browser, click Sign up, and create an account with the email you use for development work. Choose the free plan. The whole signup takes about three minutes.

Step 2. Install GitHub Desktop.

Go to desktop.github.com and download the installer for your operating system. Run it. Open the app. When it asks you to sign in, use the GitHub account you just created. The app downloads, signs in, and lands you on a screen with three big options.

Step 3. Add your project to GitHub Desktop.

Click "Add an Existing Repository from your hard drive." Pick the project folder from Chapter 4, or whichever project you're shipping. GitHub Desktop notices the folder isn't a Git repo yet and offers to create one. Click create. From now on, every change you make to files in that folder is something GitHub Desktop can see and track.

You'll notice a sidebar showing every file in your project as a "change" because Git is treating this as the very first commit. That's normal.

Step 4. Make your first commit.

In the lower left of GitHub Desktop, type a short summary into the Summary box. Something like "first commit" works. Click Commit to main. The app records a snapshot of every file in the project as your first commit. The change list clears.

Step 5. Push to GitHub.

At the top of the app, you'll see a button that says "Publish repository." Click it. The app asks for the repo name (use the same name as your project folder), an optional description, and whether to make the repo public or private. For a tool you're about to deploy to a free hosting platform, private is fine. Click Publish repository. The code uploads to GitHub. Open your GitHub profile in the browser and you'll see the new repo with all your files.

Your project is now on GitHub. From here forward, every time you change something in your project folder, GitHub Desktop sees the change in its sidebar. Type a short summary, click Commit to main, then click Push origin in the top bar. Both moves take about ten seconds combined.

(If you'd rather skip GitHub Desktop and use Git from the terminal, the equivalent commands live in any short Git tutorial. The shape is the same. Init the repo, add your files, commit, then push to a remote you set up on github.com.)

You'll come back to this loop every time you change your tool. Make a change, commit, push. Your hosting platform watches your repo and redeploys when it sees a new push.

PUSH, BUILD, LIVE

The deployment itself is simpler than everything you just decided. You push code to GitHub, the platform detects the change, runs the build, and your app goes live. The whole thing takes two to five minutes.

Connect your GitHub repo to your platform. If you're using Railway or Render for a full-stack app, the platform detects your framework, sets the build command automatically, and gives you a URL. If you're using Vercel or Netlify for a static app, the same thing happens even faster. For static sites, the whole process can finish in under a minute: sign up with GitHub, click New Project, select your repo, and the platform auto-detects your framework and deploys.

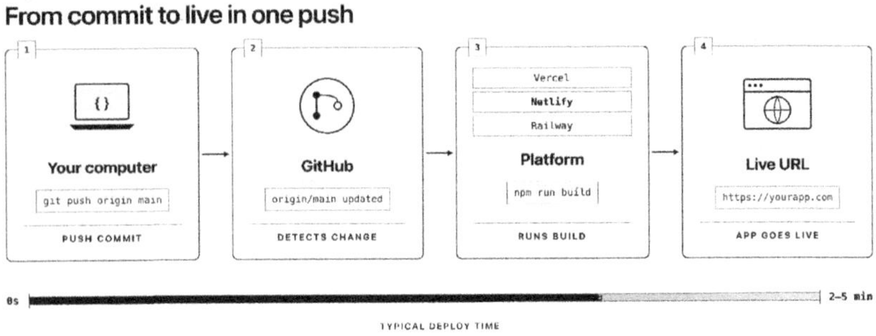

Figure 15.1: Deploying from GitHub — select your repository and the platform handles the rest.

Codecademy published a guide called "How to Deploy Your Vibe Coding Projects for Free" that walks through this exact flow.

Behind the scenes, the platform clones your repository into a clean environment. It installs your dependencies from `package.json`, runs your build command (usually `npm run build`), and serves the output. Every time you push a new commit to your main branch, this cycle repeats automatically. You push, the platform rebuilds, and the live version updates.

Reza Fatideh went from idea to a deployed multiplayer web app using Claude Code. His key insight was about treating Claude like a junior engineer who needs structure, not a magic code generator. He planned his deployment conversations before he started building, which meant that when the time came to ship, the path was already clear.

The build step is where most first deploys fail. Knowing why helps you avoid the same mistake. Your laptop has months of accumulated packages, configurations, and environment quirks. The platform's clean environment has nothing except what your `package.json` declares. Dependencies that you installed globally on your machine but never added to `package.json` won't exist. File paths that work on macOS might break on Linux because of case sensitivity. TypeScript warnings that your dev server treated as suggestions become fatal errors in a production build.

"Installed globally" means you ran `npm install -g something at some point`, which put the package on your whole computer rather than inside your project. The platform doesn't have your computer's global packages. If your code uses a package, it needs to be listed in your project's `package.json`.

Run the exact production build locally before pushing. Ask Claude to run `npm run build` (or whichever build command your project uses). If it succeeds with zero errors and zero warnings, it'll succeed on the platform. If it fails locally, it'll definitely fail remotely, and you just saved yourself a deploy cycle. The production build is a stricter teacher than the dev server. It fails on sloppiness that the dev server quietly forgives.

IF YOUR PROJECT IS PLAIN HTML, CSS, AND JAVASCRIPT

Most of this chapter assumes your project has a backend. The Chapter 4 build doesn't. If you stayed close to that build, your tool is three files (`index.html`, `styles.css`, `script.js`) plus a JSON file where submissions land. You don't have an `npm run build` step, no `process.env.X` references to wire into a hosting dashboard, and nothing like a SQLite database that needs migrating to Postgres before deploy. The deploy is shorter and simpler.

You have two paths. Pick the one that matches how comfortable you are with Git so far.

The drag-and-drop path. Open netlify.com or vercel.com in your browser and sign up with email. Look for the option that says "deploy without Git" or "drag and drop a folder." Drag your project folder onto the page. The platform builds nothing because there's nothing to build. It serves your files and gives you a live URL within sixty seconds. Each time you change something, you drag the folder over again. The trade-off is that Git history doesn't exist for this project, which means you can't roll back to a previous version through the platform. You'd need to keep your own backups.

The Git path. If you set up GitHub in the previous section, the same flow works. Push your project to a GitHub repo, connect the repo to Vercel or Netlify, and the platform serves the static files automatically. The first deploy takes about two minutes. After that, every `git push triggers` a redeploy.

The Six Things That Break section ahead still applies, but with most categories crossed off. You won't hit the missing-dependency category because there are no dependencies. You won't hit the SQLite-on-stateless-platform category because your data lives in a JSON file in your project folder. (If your tool needs to write data on the server side, like saving form submissions to a file, that's the moment your project crosses from static to full-stack. Static platforms can serve a JSON file but can't modify one.)

You'll still hit hardcoded URLs and environment variables, but in a lighter form. If your `script.js` references a file:// path that worked locally, swap it for a relative path like ./ `data.json` . If your tool calls an external API like a weather service or an email service, the API key still needs to live somewhere safe. Vercel and Netlify both let you bake static-site env vars into the build instead of running them at request time.

If you're not sure whether your project is static or full-stack, ask Claude to look at your folder and tell you which category fits. The cleanest way is a single prompt.

```
Look at this project folder. Tell me whether it's
static (HTML, CSS, JS only, no server code) or full-
stack (has a backend that runs server-side). If it's
static, tell me the simplest way to deploy it. If it's
full-stack, tell me what the backend needs.
```

Claude reads what's actually there and gives you a verdict in about thirty seconds.

THE SIX THINGS THAT BREAK

When your first deploy fails, or your app deploys but something doesn't work, the problem falls into one of six categories. They're all environmental mismatches, not code quality problems. Your code is fine. The surroundings changed.

#	What breaks	Symptom	Fix
1	Missing dependency	Build fails with "module not found"	Move the package from global install to your package.json. Run npm install <pkg> --save.
2	Environment variables missing	App starts but can't connect ("invalid key", "connection refused")	Add every .env variable to the platform's dashboard. Verify values match.
3	File-path casing	Works on Mac/ Windows, fails on Linux	Use lowercase for all folder and file names. Update imports to match exactly.
4	Hardcoded localhost URLs	Frontend can't reach backend in production	Replace http://localhost: 3000 with an env variable like process.env.API_URL.
5	SQLite on a stateless platform	Data disappears between deploys	Move to a hosted SQLite (Turso, D1) or Postgres. SQLite files don't persist on Vercel/Netlify.
6	Build command mismatch	Build works locally, fails on platform	Set the exact same build command on the platform that you use locally (usually npm run build). Run it locally once before pushing.

Table 15.2: Six things that commonly break on deployment and how to fix each one.

Environment variables are missing or wrong. Your app starts but can't connect to anything. You see "Connection refused" or "invalid key" in the logs. Open your platform's dashboard, check that every variable from your .env file has a corresponding entry, and verify the values are correct. Typos in variable names are the most common version of this, something like DATABASE_URL in your code versus

DATABSE_URL in the platform dashboard. One character off and the connection fails silently.

File path casing doesn't match. You wrote ./Components/`Header.js` in your import statement but the file is actually `./components/Header.js`. macOS and Windows don't care about the difference. Linux does, and your production server runs Linux. Use lowercase for all folder and file names. If you're using TypeScript, turn on the forceConsistentCasingInFileNames config option to catch these at compile time.

Hardcoded URLs point to localhost. Somewhere in your frontend code, there's a fetch call to http://localhost:3000/api/users. Locally, your backend is running on port 3000 and it works. In production, localhost means the user's own computer, not your server. The API call goes nowhere. Replace every hardcoded URL with an environment variable. Set it to localhost in development, set it to your production URL in the platform dashboard.

CORS blocks your requests. Your frontend is at app.yourdomain.com and your backend is at api.yourdomain.com. The browser refuses to let the frontend talk to the backend because they're on different origins. You see "Access to XMLHttpRequest blocked by CORS policy" in your browser console. This worked perfectly during local development because both frontend and backend were on localhost. CORS isn't a bug. It's browser security working correctly. Fix it on the server. Configure your backend to allow requests from your frontend's domain with the Access-Control-Allow-Origin header. The literal prompt that does this looks like the following.

```
Add the Access-Control-Allow-Origin header to my
backend so my frontend domain can call it. The frontend
lives at [paste your frontend URL here]. Use the exact
origin, not a wildcard.
```

An "origin" in browser terms is the combination of protocol, domain, and port. So https://app.yourdomain.com and https://api.yourdomain.com are different origins, even though they share the same root domain.

Never use a wildcard (*) for the allowed origin when your app uses credentials like cookies or authentication tokens. Specify the exact frontend domain instead.

Database data disappears. You deployed with SQLite, and everything you entered vanishes after every redeployment. SQLite stores your database as a single file, and stateless platforms create new instances that don't keep files between deploys. Vercel's own knowledge base says it directly. SQLite can't work on serverless platforms because the file storage is ephemeral between requests. The fix is PostgreSQL on a managed service. Railway includes it. Supabase offers a free tier. If you built with SQLite during development, you'll need to switch before deploying.

This is a one-time migration, not an ongoing headache. You just need to know about it before you deploy. Skip the warning and you'll watch your data vanish.

If your project uses a JSON file or no database at all (most plain HTML/CSS/JS readers), this category doesn't apply to you and you can move past it.

Errors happen and nobody sees them. Your app is live. A user reports that "it doesn't work." You have no idea what happened because locally, errors show up in your chat window, but production errors disappear into the void unless you've set up logging. Enable your platform's logging dashboard. Railway, Render, and Vercel all provide logs. Check them immediately after deploying and after any user reports a problem.

Most builders hit the first category, fix it, and find the next one waiting.

DEBUGGING ISSUES AFTER LAUNCH

What happens if a user emails three days after you share the URL, saying their submission failed and asking what happened? You open your tool, test the same thing they did, and it works fine for you. Now you have a problem and a user waiting.

The flow is the same iteration loop from Chapter 4, with one move added at the start. Get the bug to happen on your own machine first. Open your tool the same way the user did. If you can hit the same wall, you're in territory Chapter 4 already covered. Describe what broke, let Claude fix it, re-test.

If the bug doesn't reproduce locally, it's probably something specific to the production environment. Open your platform's dashboard. Each platform (Vercel, Netlify, Railway, Render) has a section for logs that captures every error your tool throws. Find log entries from around the time the user hit the bug. The error message tells you whether the problem is in your code, in a missing environment variable, or in something the platform itself is doing.

Once you've reproduced the bug or read the production logs, you're back where you started. Open Claude Code, describe the bug the way you would in Chapter 4's iteration loop, ask Claude to write a test that catches the failure, then fix the code. Run the full test suite. Push the fix to GitHub. Your platform redeploys automatically. Open the production URL in incognito and confirm the bug is gone.

Then write back. "Fixed in production. Thanks for catching it." Users who report bugs are doing you a favor. They're telling you about a problem other users probably hit but didn't bother to email. The faster you reply, the more of those silent users you keep.

YOUR TOOL, LIVE

You click the URL and the page loads. Not on localhost, not in a preview. On the actual internet, where anyone with the link can reach it. The tool you've been building since Chapter 4, the one you added features to, connected to external services, tested, and secured, is now accessible to anyone with the link.

Open that URL in an incognito window, not your regular browser where you might be logged in or have cached data. Click through every feature. Submit a form. Check that data persists after you close the tab and reopen it. Try what someone who has never seen your tool before would try. If something breaks, check the six categories from the previous section in order.

Once everything works in incognito, you've validated that your tool is ready for other people to use. Share the link.

A custom domain is optional, and it's less than thirty minutes of work. Buy a domain from a registrar like Namecheap or Spaceship, point it to your platform using the DNS settings they provide, and the platform provisions SSL certificates automatically. Your tool goes from yourtool.railway.app to yourtool.com in about thirty minutes.

Nomiki Petrolla is a non-technical solo founder and a mom of four who quit her job as Head of Product to build Theanna, using Claude Code to build the entire frontend. Her "Women Build Cool Sh*t" cohort now has twenty-seven women across twenty-seven states and four countries, and she joined Techstars Columbus in Fall 2025. She didn't learn to code. She learned to direct a tool that codes for her.

Vulcan Technologies went through Y Combinator's Summer 2025 batch with three co-founders, only one of them technical. They used Claude Code. Four months in, they had government contracts and an eleven million dollar seed round.

These aren't stories about people who are different from you. They're stories about people who did what you just did, chapter by chapter, from the beginning of this book to right now.

You started Chapter 1 not knowing what Claude Code was. You learned how it thinks, set up your environment, built your first tool. From there you figured out how to communicate with it, how to give it memory, how to plan before building, and how to connect your tool to real services. Then came systems that run without you, testing what you built, and locking it down. And now it's live. Someone can click a link and use the thing you made.

Appendix A: Setting Up Claude Code in the Terminal

If you'd rather use the terminal than the Desktop app, this is where to start. The install is a single command on every platform, but Windows has one prerequisite that catches people off guard, so we'll cover that first. Mac and Linux users can skip ahead to the install command for your platform.

BEFORE YOU INSTALL (WINDOWS ONLY)

For Windows users, you need Git for Windows installed first. You can download it from gitforwindows.org. It's free, and you don't need to understand git or use it yourself. Claude Code's installer relies on a component that ships with Git for Windows. Without it, the install command fails, and the error message it gives you won't tell you why.

There's one screen during the setup wizard that asks about PATH. You want the option that adds Git to your system PATH. Don't skip this step. GitHub issue #25593 documents the pain that can come with this oversight. Someone installed Git for Windows, ran the Claude Code install command, and got a cryptic "requires git-bash" error. Git was installed fine. But during the setup wizard, they'd skipped the option to add Git to the system PATH. The installer couldn't find it, so it failed. Multiple users hit the same wall with different error messages, all traced back to that one checkbox. If you already installed Git and you're hitting errors, reinstall it with that option checked.

And one thing you don't need is `Node.js`. Older guides might tell you to install `Node.js` or npm first. That was true before 2026. The native installer eliminated that dependency entirely. If someone's tutorial tells you to install `Node.js` as a prerequisite for Claude Code, the tutorial is outdated.

The installation flow looks different on each platform. Figure A.1 shows what to expect.

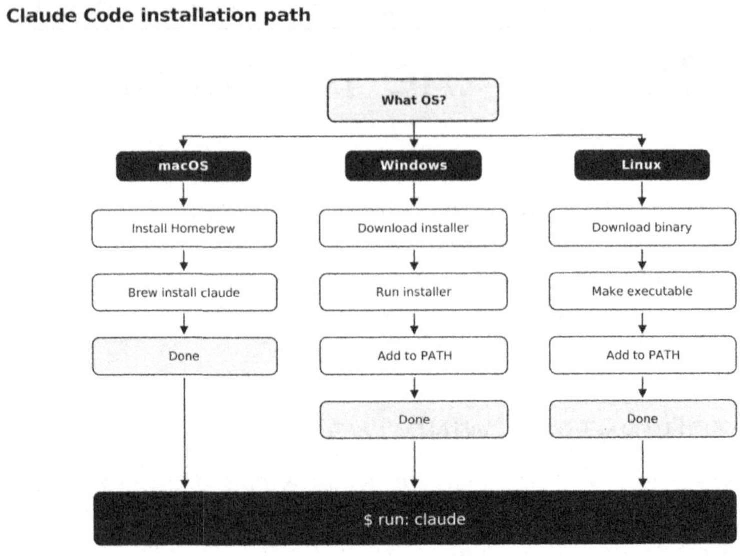

Installation paths by operating system.

Figure A.1: Installation paths by operating system.

INSTALLING CLAUDE CODE

To install, you need one command. First, open your terminal app. On a Mac, open the app called Terminal, which you can find in Applications → Utilities, or by searching for "Terminal" in Spotlight. On Windows, search for "PowerShell" in the Start menu and open it.

Once it's open, paste the install command for your platform.

On macOS or Linux:

`curl -fsSL` | bashhttps://claude.ai/install.sh

On Windows PowerShell:

irm https://claude.ai/install.ps1 | iex

Windows also has a third option. If you have WinGet installed (most Windows 10 and 11 machines do), you can use it instead. Run:

winget install Anthropic.ClaudeCode

I'd actually recommend the WinGet method on Windows, because the PowerShell command has a quirk that confuses people. You paste it, press Enter, and nothing visibly happens for fifteen to thirty seconds. No progress bar, no loading indicator, nothing. People assume it failed and run it again. The installer is working. It just doesn't show progress. The WinGet method gives you a visible progress bar and a clear success message, which is a lot less nerve-wracking when you're doing this for the first time.

VERIFYING THE INSTALL

After any of these commands finish, verify the installation by typing:

```
claude --version
```

If a version number appears, you're set. If you get an error like "command not found," close your terminal completely, open a new one, and try again. The install updates your PATH, but that change won't show up in a terminal that was already open.

FIRST LAUNCH

Now launch Claude Code by typing:

claude

Just like the Desktop app, the first launch opens your browser for authentication. Sign in, and you're done. The terminal will show a ready prompt, and you can start typing.

That gets you to a fresh prompt with no project loaded. To work on a specific project the way the body chapters describe, navigate to your project folder first, then launch a session from there. The pattern is two commands.

cd `~/my-intake-form`

claude

Replace `~/my-intake-form` with the path to whatever project you're working on. Wherever the body chapters tell you to use the Project folder control to select a folder, your terminal equivalent is cd <folder> followed by claude. Each session loads the folder you launched from, so the walkthroughs throughout the rest of the book work the same way once you're in.

Appendix B: Troubleshooting, Recovery, and Getting Unstuck

This appendix is your troubleshooting field guide. Come back to it whenever something goes wrong. If the tree doesn't resolve it, the detailed entries below walk you through the full diagnostic.

You don't need to read this front to back.

THE TROUBLESHOOTING TREE

Three root causes account for every Claude Code problem you'll hit:

Context pollution: The context window contains information working against your current task.

Instruction ambiguity: Your instructions left room for interpretation, and Claude interpreted differently than you intended.

Scope mismatch: The task is too large, too vague, or requires more context than one session can hold.

SYMPTOMS AT A GLANCE

Context pollution is the most common root cause and the hardest to recognize. The symptoms are subtle: output that's almost right but slightly off, tone that drifts mid-session, decisions that contradict ones Claude made ten minutes ago.

What you're seeing: You've been working for 15–20 minutes and the output is getting worse. Earlier responses were sharp and on-target. Now they're vague, repetitive, or drifting from your instructions.

Why it's happening: Your context window is filling up. Every prompt you send and every response Claude generates takes up space. As the window fills, Claude's ability to hold onto your earliest instructions weakens. Claude prioritizes the most recent content, pushing your original prompt and early decisions into the background.

What to do:

If the early work in this session is worth keeping, use `/compact`. This compresses your conversation history into a shorter summary that preserves key decisions and context. Your files and project folder aren't affected.

If you need a clean start, use `/clear` and write a handoff note first. Capture what you've decided, what's working, and what's still to do. Paste that note into your next session's first message.

```
I've been working on this report for a while and the output quality is dropping. Before I compact, summarize the key decisions we've made and the current state of the deliverable so nothing gets lost.
```

Claude References Things You Never Mentioned

What you're seeing: Claude mentions files from a different project, uses terminology from a topic you haven't discussed, or incorporates information that seems to come from nowhere.

Why it's happening: You launched Claude Code from the wrong directory. Claude reads every file in the folder where you start it. If you're in your Documents folder instead of your project folder, Claude pulls context from files that have nothing to do with your current task, and that pollutes the output.

What to do:

Ask Claude what folder it's in.

Close the session.

Navigate to the correct project folder.

Start a new session from there.

Claude Keeps Reverting to a Decision You Already Changed

What you're seeing: You told Claude to change the heading structure, and it did. Three prompts later, the old heading structure is back. You correct it again. It reverts again.

Why it's happening: Your one correction is fighting against all the earlier context that reinforced the original decision. If the first version of the document used one heading structure and ten subsequent prompts all referenced that structure, your one correction is competing against all of that earlier context.

What to do:

For short sessions, a direct correction might stick. Try being very explicit.

```
From this point forward, use H2 for section headers and H3 for subsections. Don't revert to the previous structure.
```

For longer sessions, `/compact` or `/clear` is more reliable. When you restart, state only your current decisions. Don't include the history of changes.

`CLAUDE.md` Instructions Are Being Ignored

What you're seeing: You wrote clear instructions in `CLAUDE.md` and Claude is ignoring them. The tone is wrong, the format doesn't match, or Claude is doing things you explicitly told it not to do.

Why it's happening:

First, check whether `CLAUDE.md` is actually loaded. Claude Code reads `CLAUDE.md` from the directory where the session starts. If you launched from a parent folder or a different project folder, your `CLAUDE.md` isn't in play. Navigate to the folder that contains your `CLAUDE.md` and start a new session.

Second, check length. If your `CLAUDE.md` has grown past 600 words, the instructions at the end get buried as your conversation grows. The fix is trimming. Cut back toward 200–400 words and front-load your most important instructions. Put the non-negotiables in the first few lines. (See Chapter 6 for the full `CLAUDE.md` design process.)

INSTRUCTION AMBIGUITY

Instruction ambiguity is deceptive because the output looks good. Claude tried hard. The writing is coherent. The structure makes sense. It just aimed at a different target than the one in your head. That gap is where all these problems start.

Wrong Deliverable Format

What you're seeing: You wanted a table and got an essay. You wanted a saved file and got a chat response. You wanted bullet points and got paragraphs.

Why it's happening: You didn't specify the output format. Claude defaults to prose paragraphs because they seem like the most natural fit for most content.

What to do: Be explicit about format in every prompt where it matters.

```
Present findings as a table with columns for date,
metric, and change.
```

```
Save the output as a file called analysis.md.
```

```
Use bullet points, not paragraphs.
```

In Claude Code, "write me a summary" gives you text in chat. "Save a summary to `summary.md`" gives you a file in your project folder. Always specify when you want a file.

```
Analyze last month's sales data. Save the results as
sales-analysis.md. Use a table for the monthly
comparison. No narrative, just the data with column
headers for month, revenue, change from previous month,
and top product.
```

Claude Added Content You Didn't Want

What you're seeing: You wanted a data summary and got recommendations, commentary, and suggested next steps piled on top of it.

Why it's happening: Claude errs on the side of thorough and helpful, which means you often get more than you asked for.

What to do: Add exclusion instructions. Exclusion instructions are just as important as inclusion instructions when you need focused output.

```
Present the data only, no recommendations, no commentary, no suggested next steps.
```

Generic Output When You Expected Specificity

What you're seeing: You asked for a client proposal and got something that could apply to any business in any industry. The output is competent but generic.

Why it's happening: Claude doesn't have enough role context. Check two things:

Is your `CLAUDE.md` loaded? If you started from a folder without one, Claude has no background on who you are or what you do.

Is the role context specific enough? "I'm a business owner" gives Claude almost nothing to work with. "I run a Shopify store selling handmade ceramics to interior designers and boutique hotel buyers" gives Claude everything it needs to make the output specific to your situation.

(See Chapter 6 for writing effective role context in `CLAUDE.md`.)

SCOPE MISMATCH

Hit scope mismatch? The fix is almost always the same. Break the task into smaller pieces. If you've arrived at this section of the troubleshooting tree, the solution is probably "stop trying to do this all at once."

Quality Degrades Across a Long Task

What you're seeing: You asked Claude to write a ten-section report. The first three sections are detailed and well-researched. Section seven is thin. Section ten reads like a different tool wrote it.

Why it's happening: Each new section fills the context window, crowding out your original instructions and earlier decisions. By section seven, Claude is holding the full text of six completed sections plus your original instructions plus all the back-and-forth. Something has to give, and it's usually the quality of new output.

What to do: Break the report into two or three sessions. Write sections one through four in session one. Create a handoff note summarizing structure, key decisions, and any patterns Claude should continue. Start a fresh session for the remaining sections.

```
Before we wrap this session, write a handoff note I can paste into the next session. Include the report structure, the tone and format decisions we've made, and a summary of what's been completed so far.
```

Claude Loops During an App Build

What you're seeing: Claude makes a change, tests it, sees an error, reverts the change, makes the same change slightly differently, and repeats. Nothing moves forward. Responses get longer as Claude "explains" its reasoning, but actual progress stalls.

Why it's happening: The build hit a complexity threshold where Claude can't hold the full architecture while reasoning about a specific change. Developer Zac from Built by Zac identified the core dynamic: Claude is pattern-matching against the error and trying plausible fixes, but when the real issue is outside the visible context, no amount of iteration will land on the answer.

What to do:

Stop pushing. Each attempt makes it worse by filling the context with failed fixes.

Write a handoff note: what works, what's broken, what you've tried, and what the intended fix was.

Start a fresh session with that note.

Consider isolating the problem: copy just the file or function that's breaking into a new folder and start a session there, so Claude only has the relevant code in context.

```
Something broke after I asked you to add the share
button to my trivia generator. The score counter
stopped updating. Here's the error I'm getting: [paste
error]. Here's what the app should do: [describe
expected behavior]. Fix the score counter without
removing the share button.
```

Claude Says It Can't Complete the Task

What you're seeing: Claude explicitly tells you it can't do what you asked.

Why it's happening: First, the task might be too vague for Claude to figure out where to start. Second, Claude might be telling you something real: the task requires a capability it doesn't have.

Claude Code can't connect to live databases during a session. It can't browse the internet in real time. It can't run software that isn't installed on your machine. If Claude says it can't do something, check whether you're asking for something genuinely outside its capability set before assuming it's a prompt problem.

What to do:

If the task is vague, make it specific. Instead of "build me an analytics dashboard," try "create an HTML file that reads data from `sales.csv` and displays a bar chart of monthly revenue."

If the task requires capabilities Claude doesn't have, rescope. Can you do the data-gathering step yourself and give Claude the results to work with? Can you break the task so Claude handles the parts it can do?

COMMON SCENARIOS

Walk through the troubleshooting tree with real situations.

Workflow Produces Wrong Output After Running Fine

Your weekly customer review analysis worked for three weeks. Week four, every review gets the same category. The workflow documentation is the same. Your `CLAUDE.md` hasn't changed. You're in the right folder.

Walk the tree:

Is the session fresh? Yes, you just started it.

Is your `CLAUDE.md` loaded? Yes, you're in the right directory.

Are the input files the same format as before? Check here.

You compare the CSV to last week's and spot the problem: the column headers changed. What used to be "review_text" is now "review_body." The platform updated its export format and Claude is reading from the wrong column.

Fix: Update your workflow documentation to specify expected column names. Add a fallback instruction to your prompts.

```
If the expected column names aren't found, list the
actual column names and ask for clarification before
proceeding.
```

One gap closed. Five minutes. You didn't rewrite the workflow. You closed a documentation gap.

Something in your setup shifted. Check the session first, `CLAUDE.md` second, inputs third. The answer is usually in the inputs.

App Build Breaks After Adding a Feature

Yesterday your trivia generator ran fine. Today you asked Claude to add a share-your-score feature, and now the score counter doesn't work. The new feature fails too.

Don't keep debugging in the same session. Each failed attempt fills the context with error messages and half-completed fixes, making the next attempt less likely to succeed.

Instead: Write down what you think the root cause is, what you've tried, and what information is missing. Start a fresh session with that diagnostic note. If the problem is isolated to one file or function, copy just that piece into a new folder and start a session there so Claude's context only contains the relevant code.

```
You just made changes to my trivia app. Walk me through
exactly what you changed, why, and what I should test
to make sure it works.
```

Claude Seems Worse Than Last Week

Quality dropped and nothing about your usage changed.

Run the environmental checklist before blaming the model:

Are you in the right folder? In the Desktop app, ask Claude Code "what folder are you working in?" and it tells you. In the terminal install (see Appendix A), run pwd.

Are the same files present as last time? Any extras that shouldn't be there?

How long has this session been running? Earlier tasks consume context space even if they're unrelated to what you're doing now.

Did you hit a rate limit? The system can quietly switch to a smaller, faster model when you've hit your usage cap.

The answer to "Claude got worse" is almost always environmental. The conditions changed more often than the tool did.

PROMPT TEMPLATES FOR COMMON SITUATIONS

Keep these in your back pocket for when things go wrong. Copy, fill in the brackets, and paste.

Debugging an Error

```
I'm getting this error when I run my [tool/app/workflow]: [paste error message]. Here's what I was trying to do: [describe the task]. What's wrong and how do I fix it?
```

Understanding What Changed

```
You just made changes to my project. Walk me through exactly what you changed, why you changed it, and what I should test to make sure it works.
```

Something Broke After a Change

```
My [specific feature] was working before I asked you to add [new feature]. Now [describe the symptom]. Undo whatever broke it without removing the new feature.
```

Starting a Diagnostic Session

```
Here's a handoff note from my previous session: [paste note]. The problem I'm trying to solve is [describe]. Start by diagnosing the root cause before making any changes.
```

Getting Unstuck

```
I've been going back and forth on this for a while and we're not making progress. Before we try anything else, summarize what we've tried, what worked, what didn't, and what you think the actual root cause is.
```

WHY TROUBLESHOOTING IS THE REAL TEST

When you diagnose a problem with Claude Code, you're working through all three of the root causes you just read about, often at once. You check whether context has filled up (the model behavior Chapter 5 explains, and the cost discipline Chapter 8 reinforces). You check whether your instructions and your `CLAUDE.md` gave Claude clear-enough direction (the prompting craft from Chapter 3 and the project-memory work in Chapter 6). And you check whether the task is matched to what one session can actually hold (the planning move from Chapter 7 and the architecture decisions from Chapter 11). And if you're really unlucky, you get all three at once.

That's the difference between someone who uses Claude Code and someone who has built a reliable working relationship with it.

For unusual errors not covered here, the Claude Code community on GitHub, Reddit, and Discord stays current as the tool evolves. What this appendix gives you is a reliable process for the problems that account for the vast majority of what goes wrong.

References

DEV Community. "Why I Stopped Copy-Pasting from ChatGPT and What I Do Now Instead." https://dev.to/teguh_coding/why-i-stopped-copy-pasting-from-chatgpt-and-what-i-do-now-instead-3d8g

CodeRabbit. "State of AI vs. Human Code Generation Report." https://www.coderabbit.ai/blog/state-of-ai-vs-human-code-generation-report

MorphLLM. "Claude Code Benchmarks." https://www.morphllm.com/claude-benchmarks

marc0.dev. "AI Coding Leaderboard." https://www.marc0.dev/en/leaderboard

DEV Community. "I Built the Same App 5 Ways: Cursor vs. Claude Code vs. Windsurf vs. Replit Agent vs. GitHub Copilot." https://dev.to/paulthedev/i-built-the-same-app-5-ways-cursor-vs-claude-code-vs-windsurf-vs-replit-agent-vs-github-copilot-50m2

X. "Anthropic Claude AI — Built with Opus 4.6 Hackathon Announcement." https://x.com/claudeai/status/2024986293248127452

adwaitx.com. "Claude Code Hackathon: Built with Opus 4.6." https://www.adwaitx.com/claude-code-hackathon-opus-4-6/

Coding Nexus. "Claude Hackathon Winners: 5 Non-Coder Apps in One Week." https://medium.com/coding-nexus/claude-hackathon-winners-5-non-coder-apps-in-one-week-736425986774

TechX. "The Day a Doctor Out-Coded 13,000 Developers at Anthropic's AI Hackathon." https://medium.com/techx-official/the-day-a-doctor-out-coded-13k-developers-at-anthropics-ai-hackathon-0d3c83e6163c

LinkedIn. "Dr. Michał Nedoszytko — Interventional Cardiologist." https://www.linkedin.com/in/michał-nedoszytko-md-phd-9243748/

ReachMD. "Michal Nedoszytko on PostVisit AI." https://reachmd.com/programs/the-convergence/Michal-Nedoszytko-PostVisit-ai/54781/

TechStory. "Cardiologist Builds Patient Care App in 7 Days, Places Third at Anthropic Hackathon." https://techstory.in/cardiologist-builds-patient-care-app-in-7-days-places-third-at-anthropic-hackathon

ICT&Health. "Doctors Can Now Turn Ideas into Working AI Tools in Days." https://icthealth.org/news/doctors-can-now-turn-ideas-into-working-ai-tools-in-days

X. "Trajektorie PL on Nedoszytko Hackathon Entry." https://x.com/trajektoriePL/status/2024774752116658539

Second Talent. "Vibe Coding Statistics." https://www.secondtalent.com/resources/vibe-coding-statistics/

Medium. "Decoding the Vibe: Insights from the r`/vibecoding` Subreddit." https://medium.com/@vtiya/decoding-the-vibe-insights-from-the-r-vibecoding-subreddit-277af7f77029

Every. "Claude Code for Absolute Beginners." https://every.to/events/cc-for-absolute-beginners

Technology Rivers. "Cursor vs. Replit vs. Claude Code: Which AI Coding Tool Should You Choose in 2026?" https://technologyrivers.com/blog/cursor-vs-replit-vs-claude-code-which-ai-coding-tool-should-you-choose-in-2026/

Matt Palmer. "Statement on CVE-2025-48757." https://mattpalmer.io/posts/statement-on-CVE-2025-48757/

NVD. "CVE-2025-48757." https://nvd.nist.gov/vuln/detail/CVE-2025-48757

Forte Labs. "The Difference Between Claude Code and Cowork." https://fortelabs.com/blog/the-difference-between-claude-code-and-cowork/

Plugins for Cowork. "Cowork vs. Claude Code." https://pluginsforcowork.com/guides/cowork-vs-claude-code/

Anthropic. "Overview." https://code.claude.com/docs/en/overview

Anthropic. "Desktop Quickstart." https://code.claude.com/docs/en/desktop-quickstart

Anthropic. "Pricing." https://claude.com/pricing

Zilliz. "Claude Code System Requirements." https://zilliz.com/ai-faq/what-are-claude-code-system-requirements

Anthropic. "Setup Guide." https://code.claude.com/docs/en/setup

git-scm.com. "Claude Code on Windows Requires Git Bash." https://git-scm.com/downloads/win

GitHub. "Claude Code Issue #25593: Git Bash Path Error on Windows." https://github.com/anthropics/claude-code/issues/25593

xda-developers.com. "Claude Code Isn't Just for Developers." https://xda-developers.com/claude-code-isnt-just-for-developers/

kevintholland.com. "Claude Code for Non-Coders: Your Terminal Is Now a Superpower." https://kevintholland.com/claude-code-for-non-coders-your-terminal-is-now-a-superpower/

Verdent. "Claude Code Pricing 2026." https://verdent.ai/guides/claude-code-pricing-2026

Anthropic. "Understanding Costs." https://code.claude.com/docs/en/costs

nxcode. "Claude Code Pricing 2026." https://nxcode.io/resources/news/claude-code-pricing-2026-free-api-costs-max-plan

laozhang.ai. "Claude Code Pro vs. Max: Eight Months of Daily Usage." https://blog.laozhang.ai/en/posts/claude-code-pro-vs-max

Anthropic. "Manage Extra Usage for Paid Claude Plans." https://support.claude.com/en/articles/12429409-manage-extra-usage-for-paid-claude-plans

karozieminski.substack.com. "Claude 1 Million Context Window Guide 2026." https://karozieminski.substack.com/p/claude-1-million-context-window-guide-2026

TrueFoundry. "Claude Code Limits Explained." https://truefoundry.com/blog/claude-code-limits-explained

Medium. "I Found a Business Idea and Shipped It in One Claude Code Session." https://medium.com/@lakshminp/i-found-a-business-idea-and-shipped-it-in-one-claude-code-session-db4ae85940d5

Medium. "From Zero to 300K Lines: How Claude Transformed Me Into a Shipper." https://medium.com/@rnnznygrxk/from-zero-to-300k-lines-how-claude-transformed-me-into-a-shipper-a73a91b3222f buildmvpfast.com. "Non-Coders Building Apps with Claude Code: Real Examples 2026." https://www.buildmvpfast.com/blog/non-coders-shipping-products-claude-code-2026

Anthropic. "Quickstart." https://code.claude.com/docs/en/quickstart

Anthropic. "Models Overview." https://platform.claude.com/docs/en/about-claude/models/overview

Anthropic. "API Pricing." https://platform.claude.com/docs/en/about-claude/pricing

kenhuangus.substack.com. "Claude Code Pattern 6: Context Management." https://kenhuangus.substack.com/p/claude-code-pattern-6-context-management

Chroma. "Context Rot." https://www.trychroma.com/research/context-rot

arXiv. "Lost in the Middle: How Language Models Use Long Contexts." https://arxiv.org/abs/2307.03172

Awesome Agents. "Long-Context Benchmarks Leaderboard." https://awesomeagents.ai/leaderboards/long-context-benchmarks-leaderboard/

DEV Community. "The Prompt Doom Loop: Why Your AI Output Gets Worse the More You Try to Fix It." https://dev.to/tejas1233/the-prompt-doom-loop-why-your-ai-output-gets-worse-the-more-you-try-to-fix-it-1j38

ProductTalk. "Vibe Coding Best Practices." https://producttalk.org/vibe-coding-best-practices/

Anthropic. "Best Practices." https://code.claude.com/docs/en/best-practices

Nathan Onn. "Claude Code Debugging: Visibility Methods." https://nathanonn.com/claude-code-debugging-visibility-methods/

HumanLayer. "Advanced Context Engineering." https://humanlayer.dev/blog/advanced-context-engineering

arXiv. "Benchmarking Prompt Engineering Techniques for Secure Code Generation." https://arxiv.org/abs/2502.06039

Stéphane Derosiaux. "From Bad Prompts to Great Code: How Claude Code Made Me a 10x Engineer." https://sderosiaux.medium.com/from-bad-prompts-to-great-code-how-claude-code-made-me-a-10x-engineer-b14d17c37f00

Theanna. "How a Non-Technical Female Founder Builds Frontend Using Claude Code." https://theanna.io/building-theanna/how-a-non-technical-female-founder-builds-frontend-using-claude-code

MDY. "With 3 Prompts, Claude Helped Me Build and Deploy a Simple Web App — No Coding Required." https://mdy.medium.com/with-3-prompts-claude-helped-me-build-and-deploy-a-simple-web-app-no-coding-required-c1fae747db86

Anthropic. "Checkpointing." https://code.claude.com/docs/en/checkpointing

Anthropic. "Interactive Mode." https://code.claude.com/docs/en/interactive-mode

Anthropic. "Keyboard Shortcuts." https://code.claude.com/docs/en/keybindings

Anthropic. "Memory and `CLAUDE.md`." https://code.claude.com/docs/en/memory

MindStudio. "Context Window Claude Code: Manage Consistent Results." https://www.mindstudio.ai/blog/context-window-claude-code-manage-consistent-results

HumanLayer. "Writing a Good `CLAUDE.md`." https://www.humanlayer.dev/blog/writing-a-good-claude-md

allarddewinter.net. "Optimising LLM Agent Instructions with `CLAUDE.md`." https://allarddewinter.net/blog/optimising-llm-agent-instructions-with-claudemd/

2nd Order Thinkers. "A Lawyer Just Beat 500 Developers." https://www.2ndorderthinkers.com/p/a-lawyer-just-beat-500-developers

AdventurePPC. "From Zero to AI Builder: How One Marketer Learned Claude Code and Automated Their Entire Workflow." https://www.adventureppc.com/blog/from-zero-to-ai-builder-how-one-marketer-learned-claude-code-and-automated-their-entire-workflow

Anthropic. "Permission Modes." https://code.claude.com/docs/en/permission-modes

Anthropic. "What's New — 2026 W13." https://code.claude.com/docs/en/whats-new/2026-w13

Anthropic. "Auto Mode." https://claude.com/blog/auto-mode

Anthropic. "Claude Code Auto Mode (Engineering Blog)." https://www.anthropic.com/engineering/claude-code-auto-mode

Wikipedia. "Armin Ronacher." https://en.wikipedia.org/wiki/Armin_Ronacher

lucumr.pocoo.org. "About Armin Ronacher." https://lucumr.pocoo.org/about/

lucumr.pocoo.org. "What Is Plan Mode?" https://lucumr.pocoo.org/2025/12/17/what-is-plan-mode/

Threads. "Boris Cherny on Spec-Driven Development." https://www.threads.com/@boris_cherny/post/DRhqRqoD6bp/

Medium. "Using Spec-Driven Development with Claude Code." https://heeki.medium.com/using-spec-driven-development-with-claude-code-4a1ebe5d9f29

Medium. "Anthropic Hackathon Results." https://kotrotsos.medium.com/anthropic-hackathon-results-b13f8466296e

X. "Cerebral Valley on Anthropic Hackathon." https://x.com/cerebral_valley/status/2026066211482857844

cc-crossbeam.vercel.app. "Crossbeam — Claude Code Community Project." https://cc-crossbeam.vercel.app/

Glama. "MCP Servers Directory." https://glama.ai/mcp/servers

workspacemcp.com. "Workspace MCP." https://workspacemcp.com

Stripe. "Stripe MCP." https://docs.stripe.com/mcp

Medium. "Building a Google Apps MCP Server for Claude Code: From Terminal to Workspace Powerhouse." https://medium.com/@mahmudulhoque/building-a-google-apps-mcp-server-for-claude-code-from-terminal-to-workspace-powerhouse-175ab40e1bf9

ArmorCode. "Predicts 2026: AI Potential and Risks Emerge in Software Engineering Technologies." https://www.armorcode.com/report/gartner-predicts-2026-ai-potential-and-risks-emerge-in-software-engineering-technologies

GitClear. "AI Assistant Code Quality 2025 Research." https://www.gitclear.com/ai_assistant_code_quality_2025_research

Ivan Turkovic. "First 1000 Lines Determine Next 100,000 in AI Coding." https://www.ivanturkovic.com/2026/02/24/first-1000-lines-determine-next-100000-ai-coding/

On Test Automation. "Writing Tests with Claude Code, Part 1: Initial Results." https://www.ontestautomation.com/writing-tests-with-claude-code-part-1-initial-results/

Google Testing Blog. "Code Coverage Best Practices." https://testing.googleblog.com/2020/08/code-coverage-best-practices.html

Autonoma. "Vibe Coding Tripled Your Bug Rate." https://www.getautonoma.com/blog/vibe-coding-quality-issues

SaaStr. "The LIVE Complete Guide to Vibe Coding Without a Developer: What We Actually Learned After Building 5 Production Apps." https://www.saastr.com/the-live-complete-guide-to-vibe-coding-without-a-developer-what-we-actually-learned-after-building-5-production-apps/

Veracode. "GenAI Code Security Report 2025." https://www.veracode.com/resources/analyst-reports/2025-genai-code-security-report/

The Hacker News. "Thousands of Public Google Cloud API Keys Exposed." https://thehackernews.com/2026/02/thousands-of-public-google-cloud-api.html

Krebs on Security. "DOGE Denizen Marko Elez Leaked API Key for xAI." https://krebsonsecurity.com/2025/07/doge-denizen-marko-elez-leaked-api-key-for-xai/

Wiz. "Exposed Moltbook Database Reveals Millions of API Tokens." https://wiz.io/blog/exposed-moltbook-database-reveals-millions-of-api-keys

SpecterOps. "Leveling Up Secure Code Reviews with Claude Code." https://specterops.io/blog/2026/03/26/leveling-up-secure-code-reviews-with-claude-code/

Anthropic. "Automate Security Reviews with Claude Code." https://claude.com/blog/automate-security-reviews-with-claude-code

Anthropic. "Automated Security Reviews in Claude Code." https://support.claude.com/en/articles/11932705-automated-security-reviews-in-claude-code

The Register. "AI Bots Hallucinate Software Packages." https://theregister.com/2024/03/28/ai_bots_hallucinate_software_packages/

BleepingComputer. "AI-Hallucinated Code Dependencies Become New Supply Chain Risk." https://bleepingcomputer.com/news/security/ai-hallucinated-code-dependencies-become-new-supply-chain-risk/

OWASP. "OWASP Top 10:2021 — A06: Vulnerable and Outdated Components." https://owasp.org/Top10/A06_2021-Vulnerable_and_Outdated_Components/

The Register. "Lovable App Vulnerabilities." https://theregister.com/2026/02/27/lovable_app_vulnerabilities/

Render. "Pricing." https://render.com/pricing

Codecademy. "Deploy Your Vibe Coding Projects." https://www.codecademy.com/article/deploy-your-vibe-coding-projects

Medium. "From Idea to Deployed App with Claude Code: What Worked and What Didn't." https://medium.com/@rfatideh/from-idea-to-deployed-app-with-claude-code-what-worked-and-what-didnt-68ac4269e96b

Vercel. "Is SQLite Supported in Vercel?" https://vercel.com/kb/guide/is-sqlite-supported-in-vercel

Y Combinator. "Vulcan Technologies." https://www.ycombinator.com/companies/vulcan-technologies

Anthropic. "Building Companies with Claude Code." https://claude.com/blog/building-companies-with-claude-code

Doppler. "Environment Variable Secrets 2026." https://www.doppler.com/blog/environment-variable-secrets-2026

MDN. "Cross-Origin Resource Sharing (CORS)." https://developer.mozilla.org/en-US/docs/Web/HTTP/CORS

Ohio Tech News. "Techstars Columbus Fall 2025 Cohort." https://www.ohiotechnews.com/techstars-columbus-fall-2025-cohort/

Anthropic. "CLI Reference." https://code.claude.com/docs/en/cli-reference

Anthropic. "Troubleshooting." https://code.claude.com/docs/en/troubleshooting

eesel AI. "A Practical Guide to Debug with Claude Code." https://www.eesel.ai/blog/debug-with-claude-code

SQLite. "When to Use SQLite." https://www.sqlite.org/whentouse.html

Addy Osmani. "The 70% Problem: Hard Truths About AI-Assisted Coding." https://addyo.substack.com/p/the-70-problem-hard-truths-about

Vercel. "Why Is My Deployed Project Showing a 404 Error?" https://vercel.com/kb/guide/why-is-my-deployed-project-giving-404

Made in the USA
Monee, IL
04 November 2026

67385772R00105